Liquid Assets

PRIVATIZING AND REGULATING
CANADA'S WATER UTILITIES

Canadian Cataloguing in Publication Data

National Library of Canada Cataloguing in Publication

Brubaker, Elizabeth, 1958-
    Liquid assests : privatizing and regulating Canada's water utilities /
Elizabeth Brubaker.

Includes index.
ISBN 0-7727-8614-3

1. Water utilities—Canada. 2. Sewage disposal—Canada.
3. Privatization—Canada. 4. Water utilities. 5. Sewage disposal.
6. Privatization. I. University of Toronto. Centre for Public Management.
    II. Title.

HD4465.C3B78 2002   363.6'1'0971   C2002-904289-5

Printed for the University of Toronto Centre for Public Management by University of Toronto Press.

TABLE OF CONTENTS

FOREWORD   3

PREFACE: TOO MANY TRAGEDIES   5

INTRODUCTION: A RISING TIDE   9

PART I: LESSONS LEARNED

Chapter 1.
Centuries of Experience: The Private Provision of Water and
Wastewater Services in the United States   17

Chapter 2.
Two Success Stories: Atlanta, Georgia, and Indianapolis, Indiana 31

Chapter 3.
The Proof in the English Pudding: Debunking the Myths About
Privatization in England and Wales   41

Chapter 4.
Eau No: The Deficiencies of the French Model of Privatization   55

PART II: WHERE'S CANADA?

Chapter 5.
A Growing Crisis: The Need for Private-Sector Involvement in
Canada's Water and Wastewater Systems   65

Chapter 6.
Testing the Waters: Early Experiments with Privatization   81

Chapter 7.
What Went Wrong? Water and Wastewater Privatization in
Hamilton, Ontario   91

PART III: MAKING PRIVATIZATION WORK

Chapter 8.
Turning Losers into Winners: Bringing Workers Onside   117

Chapter 9.
There's Nothing Like a Hanging: Creating Incentives Through
Legal Liability   129

Chapter 10.
Unnatural Monopolies: Lessons on Competition and
Economic Regulation from England and Wales   147

CONCLUSION: HOW FAR SHOULD WE GO?   159

ENDNOTES   163

INDEX   223

ACKNOWLEDGEMENTS   235

FOREWORD

In the wake of the Walkerton tragedy in Ontario – as well as water cont-amination issues elsewhere in Canada – Elizabeth Brubaker here offers the first comprehensive book-length study of privatization in the area of water and wastewater facilities. Specifically, she examines privatization experiences around the world, interpreting them with an eye to what can be learned from a Canadian perspective.

With 194 Atlantic communities pouring raw sewage into the ocean, with municipal sewage pollution closing 18 percent of Quebec's soft clam and blue mussel harvesting zones, with 92 of Ontario's sewage treatment plants failing to comply with provincial standards (according to the most recent figures, from 2000), with 121 Saskatchewan communities operat-ing deficient water treatment systems, and with 304 communities in B.C. under boil-water advisories, the Canadian status quo – public ownership and operation of water and wastewater facilities – isn't working.

On the whole, Brubaker argues, the experience elsewhere with privatiza-tion has been a good one. Private owners have invested heavily in their water facilities. Their productivity is often higher than in publicly owned utilities. Private ownership does not, contrary to what some critics argue, imply increased water exports and a consequent loss of domestic control over the resource; as Brubaker writes, "[p]rivate firms acquire the right to serve customers in a given area; they don't acquire an unlimited right to extract water from a given source for whatever purpose suits them." Efficiency gains in private water companies frequently get translated into lower rates for consumers. And where rates do go up, it's often because private companies can more easily – by comparison with politically tim-orous governments – make tough decisions to invest in costly infrastruc-ture and environmental improvements.

Perhaps most important, private water companies are actually more easy to regulate than their publicly owned counterparts. After all, when gov-ernment regulates itself it's in a conflict of interest and often pulls its punches. But as foreign experience shows, when government regulates *private* water utilities, then on many dimensions – water quality, envi-ronmental performance, customer service – the results are superior.

On top of all of this, Brubaker offers some important cautionary tales. If you're going to privatize, you can't go by half-measures. The French experience – where management of water facilities was turned over to the private sector, but where systems remained publicly subsidized and hence politicized – was less than stellar, since the managers had no incentive to undertake major capital improvements. So was the experience of Hamilton, Ontario, where management was turned over to the private sector without competitive bidding.

Brubaker notes that some union leaders, especially those who have had the opportunity to experience first-hand various kinds of public-private partnerships in water here and abroad, are becoming more open to the idea of privatization. The Canadian public, however, is not convinced. According to a 1998 poll conducted by the Canadian Union of Public Employees, Canadians, by a margin of five to one, prefer publicly to privately owned water facilities. In light of this disconnect between public perception and the actual record – and given the importance of the issue – Brubaker's book is a vital and timely entrant in the debate.

Andrew Stark
Centre for Public Management
University of Toronto

PREFACE

TOO MANY TRAGEDIES

In May 2000, E. coli and campylobacter bacteria contaminated the water system in Walkerton, Ontario, killing seven people and sickening more than 2,300. Heavy rains had washed the bacteria from cattle manure spread on a farmer's field into the aquifer serving Walkerton's Well No. 5. Once drawn from the aquifer, the contaminated water was pumped into the town's distribution system. Neither the treatment at the well-head nor the chlorine residual in the distribution system were sufficient to kill the bacteria.

Routine lab tests revealed the contamination. But the testing lab did not notify environment or public health officials of the adverse results. Instead, it telephoned Stan Koebel, manager of the Walkerton Public Utilities Commission, and faxed him the results. The bad news did not register with the distracted manager, who later claimed that the fax lay unread on his desk for three days.

As people in the community fell ill with bloody diarrhea, suspicions about water safety grew. Health, environment, and municipal officials questioned Mr. Koebel to find out if the illnesses could be traced to the water system. Mr. Koebel steadfastly denied the suggestion, repeatedly assuring people that the water was "okay." Meanwhile, he began flushing the system and raising chlorine levels. Increasingly suspicious, public health officials issued a boil-water advisory and took their own samples. Not until confronted with the results of those samples – six days after being informed by his own lab – did Mr. Koebel admit to having known of the contamination.

The public inquiry into the tragedy revealed layer upon layer of incompetence and deceit at the Walkerton PUC. Neither Stan Koebel nor his brother Frank, the foreman, understood much about water safety. The province had grandfathered both men under the operator licensing program introduced in the early 1990s. Their lack of formal training may help explain why neither knew of the threats posed by E. coli and why neither understood the details of chlorine residuals. The problems, however, went well beyond training. No training should have been necessary to discourage Frank Koebel from drinking on the job or to prevent either of the brothers from deceiving their regulators and endangering their customers by regularly misrepresenting sampling locations on water samples and recording false chlorine residuals on daily operating sheets and in annual reports.

The Koebel brothers were by no means the only weak links in the chain of source protection, water supply/treatment/distribution, and regulation. Indeed, virtually every link in the chain failed. From the family farmer whose cattle emitted the deadly E. coli to the part-time PUC commissioners who knew little about water treatment or the guidelines with which they were expected to comply, almost every party failed to take actions that could have averted the tragedy. Especially troubling was the inadequate oversight provided by the Ontario Ministry of the Environment. For years, the ministry turned a blind eye to Walkerton's problems. Although it knew that the town's shallow Well No. 5 – constructed without required government approvals – had been the subject of numerous warnings about vulnerability to contamination, it did not require the town to stop drawing water from it. Nor did the ministry act on repeated evidence of contamination in Walkerton's water system. Between 1990 and 2000, dozens of samples from the distribution system tested positive for total coliforms or E. coli. The ministry's latest inspection, in 1998, revealed not only the presence of bacteria in treated water but also inadequate chlorination, insufficient sampling, and a failure to maintain operator training records. The ministry failed to follow up on these problems.

The tragedy took a staggering toll. In addition to its immeasurable human costs were extraordinary economic costs. Economist John Livernois identified $64.5 million in hard costs – including the costs of health care, lost productivity, declining property values, replacement water, system repair, legal services, and the public inquiry – and an additional $90.8 million in the less tangible costs of illnesses suffered and lives lost.

The tragedy also cost millions of Canadians their confidence in their drinking water. It shone a spotlight on water systems all across the country, revealing inadequately trained operators, widespread non-compliance with regulations and guidelines, and an enormous backlog of unmet capital repairs and improvements. Boil-water advisories became commonplace. In June 2000, Newfoundland issued 188 advisories. Quebec followed in August with 90. Within six months, Ontario had issued 246. By August 2001, British Columbia's Health Officer reported inadequate treatment in more than 300 communities. "It is clear," he concluded, "that more can be done to . . . minimize our reliance on individual households boiling water as a *de facto* form of water treatment."

If Walkerton was, as many said, a "wake-up call," then the snooze alarm sounded 11 months later in North Battleford, Saskatchewan, alerting still somnolent Canadians to the serious problems that continue to beset drinking water systems. In April 2001, cryptosporidium contaminated the water in North Battleford, sickening as many as 8,000 people. The parallels with Walkerton were unsettling: a tainted water source (the protozoan parasite likely came from the municipal sewage plant, which periodically discharges raw sewage upstream from the town's water intake); rates inadequate to cover capital needs; malfunctioning treatment equipment; unqualified staff; operating practices so shoddy that, in the words of one municipal worker, "you're running a lot of times by the seat of your pants"; and staff's denial to public health officials that problems with the treatment plant could be behind the growing reports of diarrhea. There were also striking parallels in the lax regulatory oversight. Years earlier, Saskatchewan had discontinued its water inspection program. And although the provincial government knew that 121 rural communities had deficient water treatment systems, it had done little to correct the problems.

The sorry state of our water systems forces us to examine not only services but also institutions that we have, rightly or wrongly, long taken for granted. We must ask why our water problems, with their potentially lethal consequences, are so deeply rooted and widely spread. Why do so many utilities perform so poorly? Why do so many regulators fail to enforce laws designed to protect public health and the environment? What incentives and tools do owners, operators, and regulators need if they are to do better? What new institutional arrangements are required? Who should do what? Which parties can solve which problems most effectively and efficiently?

In searching for a better way to provide and regulate water services, one of the options we must examine is privatization. Privatization – the sale of water treatment plants or the contracting out of their operations and maintenance – has in many jurisdictions helped solve problems similar to our own. It has brought investment in infrastructure. It has made available greater expertise. It has encouraged innovation. It has promoted efficiency. It has curbed the conflicts of interest that prevent governments that own, finance, or operate water systems from strictly enforcing the laws and regulations that govern them. As a result of all of these factors, it has improved performance and brought greater compliance with health and environmental standards. Experience around the world thus suggests that privatization, accompanied by strict regulation, holds great promise for Canada.

Yet privatization remains controversial. Critics, especially those from labour unions, insist that putting water utilities in the hands of private companies will threaten the integrity of our systems, with disastrous results. Indeed, some have gone so far as to blame the Walkerton tragedy on privatization, despite the fact that the PUC was publicly owned and publicly operated. Many of the fears of those opposed to privatization are based on myths and misconceptions. This book aims to dispel some of those myths. Separating fiction from fact, it aims to demonstrate that privatization, approached properly, provides the keys to preventing future water tragedies.

INTRODUCTION

A RISING TIDE

The mid-1980s marked the beginning of a worldwide trend toward the privatization of public enterprises. Inspired by the United Kingdom – in particular, the 1984 British Telecom share issue – the governments of Denmark, Italy, Chile, Malaysia, and Singapore adopted privatization programs in 1985.[1] More than 100 countries soon followed suit, prompting a research manager for the World Bank to, in 1998, call privatization "a defining feature of the last two decades."[2]

Although estimates of the extent of privatization vary, all point to remarkable activity. A 1992 World Bank study reported that 6,800 state-owned enterprises had been privatized since 1980.[3] The Reason Foundation estimated the sale of government assets between 1984 and 1994 at US$468 billion – a figure that excluded the vast Czech and Russian assets given to citizens through voucher privatizations. The decade saw another US$83 billion worth of infrastructure projects financed, constructed, and operated by private firms working under long-term concessions. In 1996, 68 countries were planning another 664 concessions for infrastructure projects worth US$514 billion.[4] By 1999, the value of the previous 15 years of privatization projects exceeded US$900 billion.[5]

Dubbed "one of the ten mega-trends of the 90s,"[6] governments' tendency to shift functions to the private sector shows no sign of diminishing in the new millennium. In 1999, Canada's National Round Table on the Environment and the Economy anticipated private sector participation in infrastructure projects worth US$1.3 trillion over the following 10 years.[7] *Public Works Financing*, which tracks major infrastructure projects around the globe, reported in 2000 that 108 countries were considering private sector involvement in 1,371 projects with an aggregate construction value of more than US$575 billion.[8]

Popular candidates for early privatizations included telecommunications and electric power utilities.[9] Water and wastewater utilities soon followed, haltingly at first and then with greater momentum. If water was, as the *Financial Times's* John Barham suggested in 1997, "the last frontier in privatization around the world," it was a frontier that was being aggressively explored.[10] That year, *World Water and Environmental*

Engineering noted "a seemingly irreversible and rising tide of private sector involvement in the provision of water supply and sewage treatment services all around the globe."[11] By the end of 2000, at least 93 countries had partially privatized water or wastewater services or were in the process of doing so. Privatizers appeared in all regions of the world. They included local, provincial, or national governments in North America's three countries, 23 countries in Latin America and the Caribbean, 20 in Europe, 30 in Africa and the Middle East, and 17 in Asia and the Far East.[12] Private water companies now serve vast numbers of consumers. The two largest companies, Suez and Vivendi, each provide water and/or wastewater services to 110 million people.[13]

Two of the most ambitious water and wastewater utility concessions of the 1990s were negotiated on opposite sides of the globe, in Buenos Aires and Manila. Aguas Argentinas, a consortium led by Lyonnaise des Eaux, won the 30-year water and sewage concession for Buenos Aires in 1993. At the time, the deal was the largest of its kind in the world, serving a metropolitan area of 9 million people. When in 1997 the Philippines privatized Manila's water and sewage system, serving 11 million, it claimed for itself the title of the world's largest. That privatization featured the division of Manila into two zones, for which two separate 25-year concessions were awarded. One went to a consortium including the Filipino-owned Benpres and Lyonnaise des Eaux; the other went to a consortium including the Filipino-owned Ayala, U.S.-based Bechtel, and United Utilities, the parent of Britain's Northwest Water.

The privatization of water and wastewater utilities is attractive for a host of reasons that vary among countries. Pragmatism, rather than ideology, drives most privatizations. Although impressed by the successes of Margaret Thatcher's government in the United Kingdom, many subsequent privatizers do not share her conservative philosophy. Indeed, even staunchly communist governments are privatizing. Cuba has formed a joint venture with a Spanish water company to develop and operate drinking water systems for three cities over the next 25 years.[14] China has signed contracts for the construction and operation of three water supply systems and has contracted out the operation of at least 20 other water and wastewater facilities.[15] Vietnam has given two Malaysian-led consortia long-term contracts to build and operate a water pipeline and two treatment plants for Ho Chi Minh City.[16]

First and foremost, developing countries are turning to the private sector out of sheer, desperate need. More than 1.1 billion people lack access to safe drinking water and almost 2.5 billion lack adequate sanitation.[17] Sewage collected from homes is rarely treated, with 90 percent being discharged directly into lakes, rivers, and oceans. As many as 27,000 people may die each day from diseases related to water and sanitation.[18] Although estimates of the costs of providing universal access to water and sanitation facilities vary widely, they inevitably exceed current or planned public expenditures. The World Water Council estimates that annual investment must increase to US$75 billion from US$30 billion.[19]

Governments that cannot on their own finance or build the necessary infrastructure are increasingly calling on the private sector for help. Private water companies have access to enormous amounts of capital. Suez reportedly could invest almost US$8 billion a year without borrowing money.[20] It and its competitors have demonstrated their eagerness to invest in water and sewage infrastructure. Several long-term contracts, including those for Buenos Aires and Manila, have required billions in private investment.[21] Unsurprisingly, John Briscoe, senior water advisor to the World Bank, has predicted, "The long-term prognosis remains one of major private financing of infrastructure."[22] Mr. Briscoe is by no means alone. A 1998 United Nations conference on managing water supplies concluded that governments must mobilize private funds.[23] Ismail Serageldin, chair of the World Commission of Water for the 21st Century, likewise looks to private investors to bring clean and affordable drinking water to developing cities.[24]

Although governments in the developed world have greater resources and face far more manageable demands, many are nonetheless attracted to private capital. Private sector investment obviates governments' needs to borrow or to raise taxes, reducing their financial and political liabilities. It frees up public capital for competing uses. It moves financial risks away from the public purse. It is also likely to be used more efficiently than public infrastructure spending, reducing overall capital costs.

Many governments privatize to increase the effectiveness of their water and wastewater systems. Whether they are struggling to provide rudimentary service, to stem water losses – which approach 70 percent in the worst run systems[25] – or to comply with advanced health and environmental standards, they turn to firms whose many years of experience and large investments in research and development have enabled them to develop a degree of expertise rarely found in the public sector. Governments that privatize also want to improve the economic performance of their utilities. The pursuit of job creation or other social goals has left many public utilities over-staffed and inefficient. Free from public-sector practices that hinder productivity and innovation, and able to take advantage of expertise and economies of scale, the private sector enjoys greater latitude to pursue efficiencies. Disciplined by competition (increasingly, not only *for* the market but also *in* the market) and capital markets, it has powerful incentives to do so.

Privatization may correct other inefficiencies as well: those associated with the underpricing of water and wastewater services. Politicized decision making in the public sector distorts the relationship between prices and costs and encourages subsidies to various interest groups. Although endemic to public water utilities the world over, subsidies are particularly extensive in developing countries. One World Bank study found that utilities' internal cash generation accounted for only 8 percent of total project costs in Asia and 9 percent in Sub-Saharan Africa. Utilities in Latin America and the Caribbean did slightly better, generating 21 percent of total project financing, while those in the Middle East and North Africa generated 35 percent.[26] Losses of this magnitude deprive utilities of both the means and the incentives to serve more people, disproportionately hurting the poor, who remain unconnected to piped systems. Furthermore, subsidized water rates reduce consumers' incentives to conserve scarce water resources. Shifting responsibility to the private sector often allows governments to discontinue subsidies. In a fully competitive, or alternatively, a well regulated system, competition or regulation set prices that better reflect costs. Governments have fewer reasons to oppose accurate pricing – the private providers take most of the heat for price increases – and, in any case, have little authority to interfere with the markets' or regulators' decisions.

Privatization also allows for the de-politicization of environmental and health regulation. Governments that own, operate, and finance water and wastewater utilities cannot properly regulate them. All too often, conflicts of interest prevent them from enforcing compliance with laws and regulations. Privatization reduces those conflicts, freeing regulators to regulate and increasing the accountability of all parties. Enforceable contracts further increase accountability. Contracts with specific performance criteria provide governments with powerful tools to compel compliance. Contracts can guarantee water quality, maintenance levels, and capital expenditures. They can require financial assurance. And they can include financial penalties for non-compliance.

For the above reasons, governments in the developing and developed worlds alike have come to accept that their core function is to "steer rather than row."[27] Rather than owning, operating, and financing water and sewage works, they are setting policy. Rather than providing services, they are regulating them. As evidenced in the following chapters, the results of this shift have often – but not always – been impressive.

The owners, operators, funders, and regulators of Canada's water and wastewater systems have much to learn from other jurisdictions' experiments with privatization. And they have no shortage of reasons to conduct experiments of their own. Across the country, thousands of facilities fail to comply with laws and standards. Many are inefficiently run: Some are grossly overstaffed; others are staffed by insufficiently trained operators. Many are in need of costly upgrades. Water charges are insufficient to cover these costs. Clearly, many systems would benefit from the capital investment, expertise, efficiency, and accountability that privatization can bring.

The following chapters will examine a variety of asset sales, long-term leases, and short-term operating and maintenance contracts to determine what has worked, what has not worked, and why. They will focus on privatization in the United States, the United Kingdom, and France, those jurisdictions for which the most detailed information is available, whose experience is most similar to Canada's, and from whom we can learn the clearest lessons.

PART I

LESSONS LEARNED

CHAPTER 1

CENTURIES OF EXPERIENCE
The Private Provision of Water and Wastewater Services in the United States

Although new to much of the globe, the water and wastewater privatizations of the last 10 years built on a centuries-old tradition of private ownership and management in several western countries, including the United States. Private companies have supplied water to U.S. consumers since 1652, when the Water Works Company of Boston was established. At the beginning of the nineteenth century, private water companies served 94 percent of the U.S. market. Their share of the market fell as governments stepped in to service unprofitable areas. By the end of the century, their share had fallen to 47 percent.[1] It continued to fall, past 30 percent in 1910[2] to below 15 percent by 1986.[3]

A survey conducted in 1995 by the United States Environmental Protection Agency (EPA) found that 28,500 privately owned water systems served approximately 14 percent of the U.S. population.[4] These private systems were located primarily in small communities, such as trailer parks. Nonetheless, private systems did appear in many larger communities. Twelve percent of the systems serving more than 10,000 people were privately owned.[5] The EPA did not calculate the number of privately owned wastewater facilities. Its privatization coordinator estimated the number to be in the low thousands, primarily in trailer parks and small developments.[6] A 1995 study by the National Regulatory Research Institute reported that public utility commissions regulated approximately 1,300 small, privately owned wastewater systems in 28 states.[7]

In the last two decades, a handful of communities in the U.S. have sold their water or wastewater utilities to private firms. The years between 1980 and 1986 saw a dozen or so asset sales before changes to tax laws discouraged them.[8] Although tax laws have once again changed, asset sales remain rare. The highest profile sale occurred in Ohio in 1995, when the Miami Conservancy District sold the Franklin wastewater treatment plant to Wheelabrator Envirotech Operating Services. The transaction earned considerable attention, in part because the EPA chose it as a pilot project to create models for privatization.

The far more common approach to privatization in larger U.S. municipalities is the contracting out of the operation and maintenance of publicly owned water and sewage utilities. Burlingame, California, introduced this approach to the U.S. in 1972, when it contracted with Envirotech Operating Services to operate its wastewater treatment plant. Such contracts remained fairly rare in the 1980s, covering perhaps a few hundred facilities. The 1990s saw a rapid increase in their numbers. In 2000, 17 private firms surveyed by *Public Works Financing* operated 2,273 facilities for 1,882 public clients.[9] Cities contracting out water system operations now include Atlanta, Seattle, Houston, and Tampa. Those contracting out sewage system operations include Indianapolis, Milwaukee, New Orleans, and Cincinnati. Many more, including some of the country's largest cities, have been reported to be studying privatization options.

Why Privatize?

Although U.S. communities are embracing – or returning to – privately owned, financed, constructed, or operated water and wastewater systems for a variety of reasons, there is one overarching theme: Financially stressed communities with inadequate infrastructure cannot meet tough health and environmental standards on their own.

Many communities have found it difficult to comply with the Safe Drinking Water Act. In 1996, the EPA warned that one out of every five people received water from a facility that violated a national safety requirement.[10] That year, 5151 community water systems violated the EPA's maximum contaminant levels or treatment technique requirements; in addition, 15,182 community water systems violated monitoring and reporting requirements.[11] Despite progress, serious problems remain. In 2000, the EPA reported that one out of every 10 people was served by a water system reporting a health standard violation. The agency attributed at least a half-million cases of illness to microbial contamination in drinking water.[12] Many of these problems can be traced to inadequate infrastructure. Older distribution systems are deteriorating. Over one-third of the systems providing surface water need to install, replace, or upgrade filtration plants. Two-thirds of water systems need to improve storage facilities.[13]

Many communities have also failed to comply with the Clean Water Act, the major federal law governing sewage treatment. Indeed, for more than a decade following the passage of the act's predecessor, the Federal Water Pollution Control Act of 1972, violations were the norm. In 1978, the

EPA admitted that the majority of publicly owned wastewater treatment works did not meet requirements for secondary treatment. Five years later, the agency estimated that 62 percent of municipal plants still remained significantly noncompliant. The assessment of the General Accounting Office was even harsher: Complaining of "widespread, frequent, and significant" noncompliance with permit limits, it reported that 86 percent of the municipal plants randomly selected were out of compliance.[14] Although over the course of 20 years municipalities have reduced by more than one-third the amount of sewage discharged, pollution from sewage systems remains a concern. Approximately 900 cities experience combined sewer overflows, which, along with releases from sewage plants, foul beaches across the country.[15] The year 2000 saw more than 11,000 beach closures and swimming advisories.[16] The U.S. Public Interest Research Group charges that 29 percent of major municipal facilities were significantly non-compliant with their permits during at least one quarter between October 1998 and December 1999.[17] The EPA has proposed expanding permitting under the Clean Water Act to reduce sanitary sewer overflows. Since at least 40,000 overflows occur each year, such a change will likely increase non-compliance figures.[18]

The challenges presented by the Safe Drinking Water Act and the Clean Water Act have driven many privatizations. In 1998, the Hudson Institute surveyed 29 water management contracts or asset sales in 11 states. Compliance with standards was the primary driver of privatization in 34 percent of the projects surveyed and the secondary driver in another 43 percent of the projects.[19] Clearly, many communities hope, along with the EPA, that "where local governments have had difficulty meeting permit limits, privatization may result in real environmental benefits."[20]

One key factor in the private sector's ability to help communities achieve health and environmental compliance is financial. Estimates of the investments required in water and wastewater systems over 20 years vary. The EPA predicts that communities across the country will need to invest more than US$350 billion.[21] Others' estimates are much higher. The Water Infrastructure Network – comprised of state government organizations, local elected officials, environmental organizations, and industry associations – puts the cost at nearly US$1 *trillion*.[22] Current levels of investment cannot meet such needs. The Network identified an annual funding gap of US$23 billion – US$11 billion a year for water systems and US$12 billion a year for wastewater systems.

The required investments often overwhelm local governments that have to balance their budgets, are at the limit of their borrowing capacity, and fear voter opposition to tax increases. Local governments cannot count on federal assistance, since long-generous grant programs have been scaled back in recent years. As a result, many turn to the private sector for help. Bringing in private capital frees up public funds for more visible or politically popular projects. Privatization may also create income for municipalities, from the one-time windfall resulting from an asset sale, from concession fees, or from property taxes levied against private firms. Indeed, the financial attractions of privatization are so pronounced that many water and wastewater industry analysts simply assume that economic issues drive privatization. Typical are the words of privatization consultant Skip Stitt: "What's causing people to look at private management, and to a lesser degree private ownership, is simply money."[23]

Case studies confirm the role of fiscal pressures in the decision to privatize. Of the projects surveyed by the Hudson Institute, 62 percent cited financial reasons for privatizing. While half of these pointed to their need to reduce operating deficits or cut costs, the other half emphasized their need for capital investment – ranging from US$25,000 to US$250 million – as the primary driver of privatization. Another 31 percent of those surveyed cited financial issues as the secondary driver of privatization.[24] In a study of 30 water or wastewater privatizations in 16 states, prepared for the National Regulatory Research Institute, the need for funding for capital improvements tied with environmental compliance issues as the most often mentioned reason for privatization.[25]

Municipalities also privatize in hope that any capital – public or private – that is invested will be used more efficiently. Efficiencies can lower both capital and operating costs, freeing up money for other investments or reducing the rates charged to consumers. Governments are notoriously inefficient providers of water and wastewater services. Public servants have neither the tools nor the incentives to operate systems efficiently. Insufficient training has long plagued municipal wastewater systems across the U.S.[26] Constrained by rigid rules and procedures and given little discretion to operate creatively, even well trained workers can make but poor use of their knowledge. Worse, they are rarely held accountable for their actions: They are neither rewarded for increased efficiency nor punished for poor performance. These factors contribute to the "bureaucratic inertia" complained of in a report on privatization prepared for the Joint Economic Committee of Congress. That report noted that politicians also behave inefficiently, but for different reasons: "To win elections, politicians face strong incentives to confer benefits on narrow constituencies – like particular . . . subgroups of public employees – and spread the costs across all taxpayers. Concentrating benefits and dispersing costs is a tried and true formula for reelection."[27] In other words, politicians face strong incentives to operate water and wastewater systems inefficiently, particularly by wastefully increasing staffing levels.

The incentives work differently in the private sector. Competition for contracts or sales motivates bidders to reduce costs. Pat McManus, chair of the Urban Water Council of the U.S. Conference of Mayors, notes that competition, along with accountability for spending decisions, can dramatically lower capital requirements. He contends that reforms in these key areas will enable cities to meet their capital needs with just one-half of the trillion dollars estimated by the Water Infrastructure Network.[28]

Even after securing business, owners or operators cannot relax. Under many contractual or regulatory arrangements, the profit motive spurs continuing efficiencies. The compensation of staff and management alike may be tied to performance. Poorly performing individuals may be fired. As John Stokes, president and CEO of Azurix North America, explained, "People who jeopardize the company are not tolerated."[29] Poorly performing firms may also be held accountable. Their shareholders may desert them. Falling stock prices may raise the cost of capital beyond that of more successful competitors. They may face the threat of a takeover. Furthermore, a tarnished reputation may prevent them from winning other contracts. Such incentives to perform can promote a host of innovations and efficiencies. Potential savings are tremendous: The BTI Consulting Group estimates that municipal utilities could save US$25 billion a year by adopting the industry's best management practices.[30]

When developing efficient approaches to construction, operation, and management, firms call on years – in some cases, centuries – of experience. Many of the U.S.'s largest investor-owned water utilities were founded in the nineteenth century. New Jersey's Elizabethtown Water Company was founded in 1845, the San Jose Water Company in 1866, the Indianapolis Water Company in 1867, and the American Water Works Company in 1886. The world's two largest private water companies – both French firms with U.S. operations – also have long histories: Vivendi, whose U.S. subsidiary is USFilter, traces its roots to 1853, while Suez, parent of United Water, dates back to 1880. Several water companies enhance the expertise they have gained from past experience with investments in innovation. In 1999, Vivendi and Suez invested almost US$200 million between them on research and development in the water sector.[31] Not surprisingly, investments of that magnitude have led to a number of important innovations for micro-filtration, pollution detection, remote monitoring of leakage from underground water mains, metering, and flow management. The resulting knowledge of state-of-art-technological solutions is dispersed throughout a network of hundreds or even thousands of employees whose expertise can be harnessed to solve local problems. Few, if any, municipalities can match such expertise. As they face increasingly complex technical challenges, the limitations of their staff hinder their ability to meet health and environmental requirements. Many have reached the point where, in Mr. Stitt's words, they "simply can't get it done without private expertise."[32]

Although the above generalizations describe most communities' reasons for selling or contracting out the operations of their water and wastewater systems, they are by no means the only factors exerting influence. Such decisions also reflect policies emerging from federal and state bodies. Changes in tax policies have at various times raised and lowered incentives for private investment in infrastructure. In the late 1970s and early 1980s, the federal tax code encouraged private investment in infrastructure. Incentives for privatization included increased private sector access to tax-exempt debt and tax-deductible interest payments on that debt. In 1986, tax reform legislation limiting many tax incentives stalled privatization. The legislation discontinued the investment tax credit and lengthened depreciation schedules on most infrastructure. It limited contracts serving projects financed by tax-exempt bonds to five years. If a municipality signed a longer-term contract, or sold a facility, the tax-free debt would be converted to taxable debt. In such circumstances, the law also required the repayment of the depreciated value of the federal grants made in the past for the facility. In early 1997, the IRS again changed its rules, permitting management contracts of up to 20 years.[33]

Other federal policies likewise influence local privatization decisions. In 1988, the President's Commission on Privatization issued a report assessing the range of activities that could be privatized and the means of doing so.[34] Four years later, President Bush signed an Executive Order encouraging privatization to promote economic efficiency. "Private enterprise and competitively driven improvements are the foundation of our nation's economy and economic growth," the president stated in the order. "Federal financing of infrastructure assets should not act as a barrier to the achievement of economic efficiencies through additional private market financing or competitive practices, or both." The order directed federal agencies to remove regulatory impediments to the privatization of infrastructure built with federal financial assistance, to simplify requirements relating to leases or asset sales, to assist local privatization initiatives, and to increase financial incentives for privatization.[35] The Clinton administration followed suit with its National Performance Review promoting efficiency and privatization.[36] In 1994, President Clinton issued Executive Order 12893 asking agencies to "seek private sector participation in infrastructure investment and management" and to work with state and local entities to minimize barriers to privatization.[37]

Congress has sent out similar messages. In 1996, the Appropriations Committee of the House of Representatives gave the EPA the following instructions: "if qualified and experienced private sector entities can finance, build, own, operate and/or maintain wastewater treatment facilities in an equal or more cost effective manner and with the same or better environmental results, the Committee strongly urges the Agency to do everything it can administratively to remove impediments to such public/private partnerships and encourage the state and local governments to look to the private sector instead of the Federal government as a financial source of choice."[38]

The EPA has become an enthusiastic proponent of privatization. It established a Public-Private Partnerships Initiative, later renamed Partners Rebuilding America, to encourage municipalities to meet infrastructure needs with private financing. To assist municipalities in privatizing, the agency publishes case studies of privatization successes, a "self-help guide" for municipalities that wish to engage in public-private partnerships, and other materials providing guidance on financing options, partnership arrangements, and contract development.

A number of states have likewise pushed privatization. One of the more interesting examples is Georgia, which passed House Bill 1163 in 1998. The law requires the owners of large public sewage treatment plants that repeatedly breach pollution limits to contract out their operations and maintenance. It applies to facilities that exceed monthly limits for biochemical oxygen demand, suspended solids, ammonia, or phosphorus in eight of 12 months, to those that significantly exceed such limits in four of 12 months, or to those that experience three major facility bypasses in one year. The owners of the offending plants must select contractors through competitive bidding processes and enter into contracts for the operations and maintenance of the sewage collection and treatment systems for between 10 and 50 years. Those failing to comply with the legislation may face multiple fines of US$50,000 and US$100,000 per day. There remains one category of reasons that U.S. municipalities give for privatizing their water and wastewater systems. Under the loose heading of depoliticization, this category includes goals as disparate as increasing accountability through clearly stated contractual obligations, bringing the pricing of water and wastewater services – which are notoriously underpriced and overused in the U.S. – in line with costs, and limiting government. The latter is especially common. In interviews with 117 officials, the General Accounting Office learned that governments are privatizing in large part to reduce the scope and size of government.[39] Since the private sector has demonstrated its ability to supply water and wastewater services efficiently and effectively, many communities feel that there is no obvious need to do so themselves. Communities are increasingly agreeing with New York Governor Mario Cuomo's statement that "the purpose of government is to make sure services are provided, not necessarily to provide services."[40]

The Results

Privatization in the U.S. has, on the whole, lived up to its promise, both financially and environmentally. Water companies have invested considerable sums in infrastructure. A National Association of Water Companies survey of 84 investor-owned water utilities serving 5.7 million households and businesses found that the firms had invested almost US$983 million in 1998 and planned further capital expenditures of almost US$4.2 billion in the following five years.[41] The Hudson Institute's survey found that private firms that purchased or leased facilities invested far more than those that merely operated facilities. The 16 firms involved in operations and maintenance contracts or outsourcing agreements made no significant capital expenditures. In contrast, the nine firms that purchased assets invested US$38 million in new or upgraded facilities and equipment (exclusive of acquisition costs), while the four involved in long-term leases invested US$18 million.[42] Nonetheless, some operations and maintenance contracts do involve large investments. United Water has invested almost US$10 million in advanced technologies for Atlanta's drinking water system.[43] The firm invested more than US$3 million in the first six years of its contract with Hoboken, New Jersey, and expects to invest another US$3.5 million in the coming 10 years.[44]

A legion of efficiencies has brought tremendous savings from privatization. By streamlining finance, design and engineering, procurement, and construction practices, private firms have reduced construction times and costs. Free from political constraints, they have cut staffing levels. They have invested in costly equipment promising long-term savings. They have developed innovative management information systems and data processing technologies to improve cash flows, accounting, metering, billing, and debt collection. Large firms have taken advantage of bulk prices for chemicals and other supplies and have benefited from economies of scale in design, expertise, and equipment. The resulting savings have been, in the words of Mr. Stitt, "just mind boggling."[45]

Numerous studies give credence to such enthusiasm.[46] Trent University economist Harry Kitchen reviewed three U.S. studies from the late 1970s. One study of 112 water suppliers found public firms to be 40 percent less productive than their private counterparts. When one of the public suppliers became private, the output per employee increased by 25 percent. Conversely, when one of the private suppliers became public, the output per employee declined by 40 percent. A second study of 143 water suppliers found costs to be 15 percent higher for public firms. A third study found public modes to be 20 percent more expensive.[47]

The Reason Foundation has repeatedly found private firms to be considerably more efficient than their public counterparts. A 1992 study concluded that contracting out water services achieved operating cost savings of between 20 and 50 percent. Examples included 40 percent savings on wastewater treatment in New Orleans and 30 percent savings on wastewater treatment in Schenectady.[48] In a 1996 study comparing the performance of 10 government-owned California water companies with that of the state's three largest investor-owned water companies, the Reason Foundation calculated that annual operating expenses per connection averaged US$330 for the former and US$273 for the latter. Proportionally, the government-owned companies hired more than twice as many employees and spent almost three times as much of their operating revenues on salaries. Furthermore, they spent almost twice as much on maintenance to produce a product of the same quality.[49] When subsidies were accounted for, water from public operations cost 28 percent more than water from private operations.[50] Another Reason report documented savings in other jurisdictions, including those of 43 percent from the competitive contracting of operations of a water purification plant in Houston.[51]

Public Works Financing's estimates of the operating savings resulting from outsourcing, based on 45 operations and maintenance contracts with terms of over 10 years, fall in roughly the same range as those reported above: 20 to 45 percent. The magazine reports that Tampa Bay's 20-year contract for the design, construction, and operation of a surface water treatment plant will bring savings of 21 percent. Newport's 20-year wastewater management contract was priced at 24 percent below the city's base cost. Milwaukee's 10-year wastewater treatment contract guarantees savings of 30 percent. New Haven's 15-year wastewater contract will likewise bring savings of 30 percent. In Seattle, the contract to design, build, and operate the Tolt River water filtration plant for 15 years was priced at 40 percent below the city's benchmark. Atlanta's 20-year contract to operate and manage its drinking water system will save the city 44 percent. Savings promise to be even higher in the future: In Tampa Bay, the cost of a 30-year contract to design, build, and operate a seawater desalination plant is expected to be just half the public benchmark.[52]

Other examples of the magnitude of savings from privatization abound. The privately designed, built, and operated water treatment plant for New Jersey's Howell Township cost 25 percent less than similar plants in the area; operating costs at the extensively automated plant are also lower.[53] In Mount Vernon, Illinois, the private expansion and operation of the wastewater treatment plant – undertaken in order to bring the plant into compliance with environmental regulations – not only saved the city 32 percent but also solved the problem six years earlier than the city could have done.[54]

A representative of one large water company suspects that some of the most dramatic savings mentioned above represent losses for the bidders. On the subject of Atlanta, for example, he commented, "It all boils down to who wanted to lose the most money for the longest time." To the extent that firms so highly value an opportunity to establish themselves that they are willing to underbid their competitors at a loss to themselves, savings will be more modest in the future. Nonetheless, even this cautious insider estimates savings of between 10 and 30 percent – with an average range of 10 to 20 percent – for both capital works and operations and maintenance.[55]

In addition to providing cost savings, privatization has frequently brought significant infusions of cash into municipal coffers. The nine asset sales examined by the Hudson Institute had a price tag of US$537 million. Lease agreements and operations and maintenance contracts may also provide cash up front. The Hudson Institute reviewed six concession fees totalling US$35 million.[56] Municipalities have experimented with other arrangements as well. As part of its 25-year wastewater treat-ment plant lease, Cranston, Rhode Island, negotiated an upfront lease payment of US$48 million, which it used to repay debt and to establish a working capital fund to enhance the city's credit standing. The city expects other benefits of privatization to include upgrades to its facility, savings of US$96 million, and lower user fees.[57] The US$10-million con-cession fee to Wildwood, New Jersey, will be paid in stages over the first 10 years of the 20-year service agreement.[58] Some contractors have com-bined one-time concession fees with additional annual payments. Scranton, Pennsylvania, negotiated a US$8 million concession payment and US$620,000 in annual fees over four years from its five-year waste-water contract, while in a 20-year contract for its water treatment plant, Rahway, New Jersey, negotiated US$13 million in concession fees over three years, followed by annual payments.[59]

The variations have seemed endless. Brockton, Massachusetts, was under considerable financial strain when it decided to contract out the opera-tion, maintenance, and management of its water and wastewater plants. Facing a budget deficit of over US$10 million and restrained by state law from raising property taxes, the city wanted a spending breather. It nego-tiated a contract that allowed it to realize all of its savings from privati-zation up front, paying nothing for treatment in the first year. Nine years into the contract, the public works commissioner expressed his delight with the plant's performance, saying that "there has never been a day where we had to worry."[60]

The financial savings from privatization have sometimes – but by no means always – translated into lower rates for consumers. In the first four years following the 1995 sale of the Franklin, Ohio, wastewater treatment plant, rates fell by 14 percent. The EPA, comparing projected rather than actual rates, credits that privatization with a 28 percent reduction.[61] Milwaukee's wastewater contract enabled the city to cut sewer fees by 15.5 percent after one year.[62] In other cases, although rates have not fall-en, neither have planned rate increases materialized. Through a 20-year water contract, Atlanta pared a projected 100 percent rate increase down to 30 percent.[63] The Hudson Institute found that privatization enabled several communities to completely eliminate planned rate increases of 5, 35, and 50 percent. Other communities tempered rate increases signifi-cantly: One reduced a planned increase from 100 percent to annual increases of between 3 and 5 percent, while another reduced a planned increase from 50 percent to annual increases of between 3 and 10 per-cent.[64] In contrast, the Reason Foundation's study of California utilities did not find rates to be lower for privatized water services. It concluded, however, that even though private companies must pay taxes – averaging US$41 per connection – that public companies avoid, and even though they do not enjoy public subsidies, their greater efficiencies enable them to offer comparable services at comparable prices.[65]

Communities have also benefited from the cleaner environment that often follows privatization. The private sector's capital investments have paid off in improved environmental performance. Of the facilities surveyed by the Hudson Institute, 12 had been out of compliance with environmental regulations at the time of privatization. Within one year, all had achieved full compliance.[66] Improved compliance has not resulted from an infusion of private capital alone; it has also reflected private-sector expertise. The new owner of Franklin, Ohio's, plant attributes the drop in violations of permit requirements – from 30-40 per year to one – not only to upgrades of the plant's aeration systems but also to more efficient plant operation.[67] Some communities have given their contractors financial incentives to improve environmental performance. Milwaukee established a system of performance payments and penalties related to the quality of effluents from its two wastewater treatment plants. For example, it rewards the contractor for reductions in annual average biochemical oxygen demand, adding US$100,000 to the contractor's service fee for every milligram per litre of improvement. For its first year of operations, the contractor earned a US$50,000 bonus along with kudos for consistently meeting national permit requirements for the first time in five years. It repeated this performance the following years.[68]

Of course, not all privatizations have been successful. Charges of corruption have tainted a handful of contracts. In 1996, the FBI launched an investigation into contracting irregularities involving Professional Services Group (PSG) and its parent, Aqua Alliance. Although it initially investigated the bidding for a wastewater plant in Houston, it expanded its focus to include other municipalities. (PSG has since been acquired by Vivendi and folded into USFilter; charges resulting from the investigation have been against employees of or consultants to PSG or Aqua Alliance rather than Vivendi or USFilter.) In 2001, Aqua Alliance pleaded guilty to bribing the chairman of New Orleans's water and sewer board and his business partner while PSG was operating the city's sewage treatment plants in the mid-1990s. It agreed to pay a US$3-million fine. In a separate action, four people affiliated with PSG and a former New Orleans water board member were indicted on conspiracy, bribery, and fraud charges. PSG allegedly conspired to provide cash and free legal services to the water board member in exchange for her help in extending its wastewater contract. The year also exposed PSG's corrupt activities in other communities. In Bridgeport, Connecticut, two close associates of the mayor pleaded guilty to bribery, fraud, and tax evasion in connection with the wastewater management contract awarded to PSG in 1997.[69] And in Jeanerette, Louisiana, a PSG employee was convicted of attempting to bribe an alderman with the offer of a job in exchange for his vote to award a water management contract to the firm. The former mayor was also charged in connection with the matter.[70]

Despite the probe into PSG, there is no evidence to suggest that corruption is a widespread problem. Nonetheless, privatizers cannot simply assume that their processes and projects will be clean. Given the risk that even limited corruption may interfere with competition, deter investors, compromise efficiencies, increase costs, reduce revenues, and undermine the public's confidence in institutions, the issue should not be ignored. Governments can ensure the integrity of the privatization process by restricting post-government employment with private contractors, limiting corporate contributions to political campaigns, insisting upon competitive bidding for contracts, and creating transparent processes and institutions from which information is readily accessible. Transparency both deters corruption and allows the public to detect any that does occur. It enables citizens to hold accountable those entrusted with power and increases the likelihood that they will reap the full benefits of privatization.

Some communities, dissatisfied with private owners or operators, have recently "municipalized" water and sewage works. Most have done so in order to more directly control rates and rate design, to control water resources, to achieve tax and financing advantages, to keep profits to spend elsewhere, or to shift to contracting out.[71] Indianapolis purchased its local water company in April 2002, explaining that a recent merger had required the company's parent to divest all water utility assets and citing concerns that a sale to another private company could have brought significant rate increases.[72] But the city had no intention of operating the utility itself. Four months earlier, it had initiated a competitive bidding process that resulted in a 20-year contract with USFilter to operate and maintain the drinking water system, to manage capital improvements, and to provide customer service.[73]

Despite several well-publicized examples (many of which, like the Indianapolis example, are somewhat ambiguous), municipalization remains the exception to the rule. Most communities appear to be satisfied with the results of their privatizations. *Public Works Financing* examined the 1998 record on contract renewals for 13 of the U.S.'s largest water contract operators. Of the 127 contracts that expired in 1998, 107 contracts – 84 percent – were renewed with the incumbent operator. Ten contracts were negotiated with different private operators. Two contracts fell into an unexplained "other" category. Eight contracts – 6 percent – were re-assumed by municipalities.[74] The magazine repeated the survey two years later, with similar results. Of the 166 contracts that expired in 2000, 151 contracts – 91 percent – were renewed with the incumbent operator. Six were negotiated with private competitors. Nine – just 5 percent – were re-assumed by municipalities.[75]

CHAPTER 2

TWO SUCCESS STORIES
Atlanta, Georgia, and Indianapolis, Indiana

Water Utility Privatization in Atlanta, Georgia

Atlanta privatized its drinking water system for one overarching reason: Privatization would dramatically reduce annual operating costs. The city's water and, especially, wastewater systems required costly upgrades, and unhappy consumers faced significant increases in their water and sewer rates. Privatization would free up money for repairs while moderating rate increases. It provided an environmental, financial, and political lifeline.

Throughout much of the 1990s, Atlanta's wastewater system caused nightmares for residents and politicians alike. Aging sewers and inadequate treatment plants contaminated land and water. Twenty-foot-high "faecal fountains" periodically erupted from man holes. Overflows, spills, and leaks contaminated the Chattahoochee river system with raw sewage. Federal, state, and private complaints resulted in millions of dollars in fines and a consent decree specifying expensive corrective action.

Less severe problems beset the water system. Water main breaks were common in the winter. A 1994 rupture dramatically reduced water levels, necessitating extreme conservation measures.[1] In 1997, state inspectors cited the city for problems with record keeping, monitoring, staffing, and discharges of filter backwash water.[2] One privatization consultant described the system's management as "very political" and, perhaps as a result, "not terribly efficiency-conscious."[3] A local newspaper called the improperly functioning system "a dangerous embarrassment" and urged the city to "get the system out of bureaucrats' hands and into those of specialists who know what they're doing."[4]

Mayor Bill Campbell initially rejected privatization as a way of solving the system's problems. Indeed, the populist democrat opposed the privatization of city services when he was running for office.[5] Reflecting on his subsequent change of heart, he admitted to one reporter, "It's an odd circumstance, because I don't favour privatization philosophically."[6] To another reporter he explained, "Privatization is a government's admission of failure. Government ought to be able to accomplish projects as efficiently as business."[7] Apparently, grim economic realities swayed him. In an apologia explaining his decision to privatize, the mayor noted that the city's water and sewage systems needed almost US$1billion for immediate improvements and that, in the absence of privatization, average rates would increase immediately by over 50 percent.[8] Such increases, he said, would place "an unbearable financial burden" on ratepayers; for senior citizens and low-income residents, they would be "unacceptable, not to mention immoral." Privatization could generate needed funds without undue rate increases. Residents would benefit, the environment would benefit, and the city could focus its attention and energy on other pressing needs. "Privatization," the mayor wrote, "is one strategy whose time has come . . . . There simply is no viable alternative."[9]

Having explored its privatization options, and having rejected an asset sale as requiring more political discussion than its time frame permitted,[10] the city decided to proceed with a contract operations arrangement. At a cost of over US$2 million, it hired a number of engineering, financial, legal, and environmental consulting firms to help it design, execute, review, and revise the process.[11] Competitive selection commenced in 1998. Five firms responded to the city's request for qualifications, and all five went on to submit proposals, although only four completed the process. In evaluating the proposals, the city and its consultants used a point system that weighed, in declining importance, annual costs, technical and management quality, minority participation, employee relations, and experience.[12]

Opposition to the process came from several corners. City council member Clair Muller objected that the privatization process was moving forward too quickly and expressed concern that a 15- or 20-year contract would create a monopoly. Even so, she supported the idea of privatization.[13] Louder criticism came from the Metro Group, a group of current and former business and government leaders that opposed the commitments, in the proposed privatization agreement, to affirmative action and equal opportunity.[14] This group and other critics also maintained that the mayor's handling of privatization was prone to corruption.[15] Concerns were raised both about the relationship of one of the city's privatization consultants to a firm bidding on the project and about contributions made by that firm to Mr. Campbell's mayoralty campaign – concerns not put to rest until the firm did not win the contract.[16] Despite such opposition, on the whole, privatization received accolades from local newspapers, council members, and business coalitions.[17]

In October 1998, with two dissenting votes, city council approved the selection of United Water Services Atlanta (UWSA), a partnership between United Water Services and Williams-Russell and Johnson, a local engineering firm. The partnership offered the lowest cost: At US$21.4 million, its annual cost for a 20-year contract was between US$1.3 million and US$4.5 million below that of its competitors.[18] Furthermore, it offered a good track record. United Water boasted many years of experience: It was founded in 1869, and its part-owner, the international water giant Suez Lyonnaise des Eaux, was established in 1858.[19] It could cite several notable privatization successes, even when measured in terms of labour relations. Indeed, the firm's good labour relations in Indianapolis and Milwaukee helped influence the choice of contractors.[20]

UWSA's bid was also attractive for its socio-economic promises. The minority-owned Williams-Russell and Johnson provided 35 percent minority participation in the business – an important factor in the city's decision.[21] Furthermore, it offered benefits to the city's poorer neighbourhoods, which were the mayor's power base.[22] In order to boost economic development in Atlanta's inner city "Empowerment Zone," UWSA agreed to locate its regional headquarters there and to encourage its employees to relocate there. It also committed to hiring 20 percent of its workforce from the zone, to helping companies start-up operations in the zone, and to providing US$1 million in annual funding for water research at the zone's Clark Atlanta University.[23] The firm has received kudos – from the city and, in 2001, from the U.S. Conference of Mayors – for its efforts to provide opportunities for the community's less privileged residents and businesses.[24] The firm itself has benefited from tax incentives offered through the Atlanta Empowerment Zone Corporation. Indeed, it has been suggested that anticipated tax incentives of up to US$8,000 per employee contributed to UWSA's low bid price.[25]

The parties signed a twenty-year agreement that went into effect on January 1, 1999. The agreement covered the operations and maintenance of two water treatment plants serving 1.5 million people in the greater Atlanta area – an area covering 650 square miles. It also assigned to the company responsibility for 12 storage tanks, 7 pumping stations, 25,000 fire hydrants, 2,400 miles of water mains, billing, collections, and customer service.[26] Although Atlanta retained responsibility for most capital investments, UWSA agreed to invest almost US$10 million in automation and information technologies.[27]

The contract set UWSA's annual operations and maintenance fee at

US$21.4 million – 44 percent less than the US$49 million the city had previously spent running the system. Some costs remained with the city: It would spend approximately US$6 million annually on power, insurance, and contract-monitoring.[28] Regardless, with 20 years of savings of between US$20 million and US$30 million a year, the city would be guaranteed savings of US$400 million over the life of the contract. Mayor Campbell vowed to use those savings to repair the water and sewer systems.[29]

One source of savings for UWSA was the reduction in staff made possible by cross-training, increased employee productivity, and computerization. The city's request for proposals had prohibited layoffs in the first three years of private operations.[30] UWSA went further, guaranteeing no layoffs for the life of the contract.[31] Regardless, many staff members left voluntarily. When the deal was approved in October 1998, the water department had 535 employees.[32] By the time UWSA took over, that number had declined to 479. All 479 were offered jobs with current wages and benefits; 417 accepted.[33] Another 67 workers left UWSA during the first year of contract operations.[34]

In several important respects, labour gained ground in the privatization. The city had not previously certified the union, which had no collective bargaining agreement. Several months after signing its contract with the city, UWSA signed a three-year agreement with Local 1644 of the American Federation of State, County, and Municipal Employees. It was the first agreement in Georgia between a private firm and a public sector union. The agreement provided union members – many of whom had not had a raise in several years – with a signing bonus, wage increases, and bonuses for improvements in efficiency.[35] UWSA compensated for the decrease in some benefits with more generous pension contributions and better medical plans for the workers. Even so, not all workers were happy with the change. Concerned about the loss of health benefits after retirement, 250 employees tried unsuccessfully to block privatization in the courts.[36] Larry Wallace, the city's Chief Operating Officer, noted that some workers' "adverse attitudes" made the transition difficult.[37]

Several other challenges greeted UWSA in its new job. The firm inherited from the city between 4,000 and 7,000 outstanding requests for service, some of which were three years old. (The number varies in the city's and firm's estimates.) The backlog prevented the firm from responding to leaks within one day or installing meters within 15 days – performance requirements that would kick in later in the contract period – leaving customers, in the words of one columnist, "fed up over leaky pipes and lengthy repairs."[38] The firm tackled the problem by installing a computer system to track work orders and cross-training workers to enable them to repair pipes more efficiently.[39]

A consultants' audit of UWSA's first-year performance noted that both the backlog of work and the unanticipated volume of new work resulting from growth in the service area contributed to some problems maintaining the system. It charged that the loss of senior staff and the firm's failure to promptly secure additional experienced repair crews exacerbated these start-up problems. Nonetheless, the audit was generally favourable. It concluded that all of the firm's charges were warranted and showed no signs of being manipulated to increase revenues, that the firm minimized costs to the city where possible, and that the firm used state-of-the-art technology and environmentally sound practices.[40]

In terms of water quality, performance neither dramatically improved nor worsened. Atlanta's drinking water met or surpassed all state and federal standards in the years immediately preceding and following privatization. Some contaminant levels decreased during the first year of private operations while others increased. The measure of total coliform bacteria fell from 1.4 percent in 1998 to 0.8 percent in 1999. Levels of copper increased from 180 to 200 ppb, lead from 4.1 to 5.5 ppb, nitrate as nitrogen from 0.5 to 0.6 ppm, and total trihalomethanes from 46 to 47.1 ppb. Trihalomethanes declined significantly the following year. Sampling increased dramatically. In 1998, the city collected over 2,000 samples and conducted over 10,000 tests. In 1999, those numbers rose to 12,000 and 50,000, respectively, and remained there the following year.[41]

In 1999, Mayor Campbell professed his administration "extremely pleased with our transition in this public-private partnership."[42] The following year, he announced his intention – which would remain unrealized – to build on the success of water privatization by privatizing the city's sewer system.[43] The mayor's enthusiasm persisted. In 2001, he boasted of spearheading the "bold" arrangement that brought savings, inner-city investment, and service improvements.[44]

Shirley Franklin's assumption of the mayorship in January 2002 marked the beginning of a more difficult phase in the city's relationship with its water contractor. Mayor Franklin, who had worked with USFilter on its unsuccessful 1998 bid for the water contract, was not philosophically opposed to privatization. She was concerned, however, about UWSA's slowness to collect bills, its backlog of meter installations, and its inadequately trained maintenance staff.[45] The mayor gave the firm until mid-November to correct the deficiencies.

UWSA defended its record, maintaining that its performance surpassed that of the city. It blamed insufficient collections on delinquent public customers. It attributed delays in installing meters to a surge in demand following a construction boom. It observed that its workload had greatly exceeded the city's initial specifications. And it pointed to accomplishments, including reductions in water losses, improvements in equipment maintenance, and improvements in employee safety.[46]

The spring of 2002 also brought disputes over water quality. Discoloured water flowed from many taps, and several boil-water advisories were issued. Whether or not the problems reflected circumstances beyond UWSA's control, they tried the public's confidence in its water system.

While Atlanta and UWSA worked to resolve their concerns, privatization continued apace, reflecting undiminished confidence in the process. Fulton County, in which Atlanta largely sits, signed two deals privatizing sewage operations. OMI won a contract to operate and maintain three sewage plants and 29 pumping stations in April 2000.[47] Seven months later, the county awarded to Azurix a contract to design, expand, and operate a fourth sewage plant.[48]

Wastewater Utility Privatization in Indianapolis, Indiana

Privatization in Indianapolis has been an unqualified success.[49] The city contracted out the operations and maintenance of its two sewage treatment plants in 1994. Two years later, it contracted out the operations and maintenance of its sewage collection system. Combined, the contracts will save the city more than US$250 million by 2007.[50] Privatization has also enhanced environmental performance and improved relations with the systems' employees.

Privatization of the sewage treatment plants began with a competitive bidding process in which all documents, including the contract, became public. From a field of five bids, the city selected the White River Environmental Partnership (WREP), a consortium including IWC Resources (parent to the company that had supplied drinking water to Indianapolis for over 100 years), JMM Operational Services (now United Water Services), and Suez Lyonnaise des Eaux. WREP brought extensive expertise to Indianapolis.

According to Mike Stayton, director of the city's public works department, "It's just a different league. These guys have resources our guys could only dream of." Mayor Stephen Goldsmith added, "WREP brought us some of the best technical experience in the world – the companies comprising the partnership employ more PhD civil engineers than the city of Indianapolis has employees. They literally wrote the book on water treatment."[51]

Its appreciation of the contractor's experience had not prepared the city for cost savings of the magnitude that occurred. Before privatization, consultants Ernst and Young had estimated that contracting out operations of the recently renovated and apparently well run sewage treatment plants would achieve savings of just 5 percent.[52] That estimate was off by a factor of eight. The city had spent US$30.1 million on its plants in 1993; in the first year of private operations, costs fell to US$17.6 million – a drop of almost 42 percent.[53] Over the first five years of the contract, the city saved US$72.8 million, including US$63.1 million in operations and maintenance costs and US$9.7 million in avoided capital expenditures.[54] These savings were US$7.9 million greater than promised in WREP's initial proposal.

In part, the savings reflect lower staffing costs. Rather than specifying staffing levels, the sewage treatment contract simply requires WREP to employ adequate staff to operate the plants at specified performance levels.[55] This flexibility allowed WREP to reduce staff from 322[56] to 196 at the start of the contract.[57] Four years later, just 157 people remained at the plants – fewer than half the pre-privatization number.[58] WREP's decision to use chlorine rather than ozone to disinfect the plants' effluents brought further savings.[59] Savings also resulted from the use of technology that was previously unavailable to the city, economies of scale allowing for wholesale purchase agreements, and improved planning, including greater emphasis on preventative rather than corrective maintenance. In its first year, WREP decreased inventory from US$6.7 million to less than US$2 million and reduced the number of warehouses from 37 to two.[60] By the end of the second year of private operations, utility costs had fallen by 20 percent, corrective maintenance costs had fallen by 30 percent, and unanticipated capital expenditures had fallen by 20 percent.[61]

Cost savings have enabled Indianapolis to keep taxes low, prompting Mayor Goldsmith to applaud privatization's "enormous benefits for taxpayers." Savings have also enabled the city to avoid raising sewer rates, which remain at 1985 levels. The city has invested much of the savings in repairing its crumbling sewer system. According to one source, the city's investment in sewers has totalled US$30 million.[62]

Elsewhere, Mayor Goldsmith has been quoted as saying that privatization allowed the city to invest more than US$90 million in re-building the sewer system.[63]

Under a differently structured contract, cost savings might have been even greater. While the contract obliges WREP to fund routine maintenance and operations, the city retains responsibility for all major investments. It will fund up to US$3.5 million a year in corrective maintenance and minor capital improvements and another US$3 million a year in major capital improvements. The city's agreement to reimburse WREP for labour, materials, and subcontracting at cost plus 11 percent mark-up may curb WREP's incentive to reduce the costs of major maintenance and capital improvements.[64]

Labour relations

Despite extensive cuts to staff, relations between management and labour have improved. Thanks to "job banking" within city government operations (a process that involved leaving positions open in anticipation of an influx of transferred workers, rather than filling them with new employees) and an extensive outplacement program funded by WREP, the transition from public to private operations left no workers unemployed: Sixty-seven found positions with the city; 43 found private sector jobs through the outplacement program; 10 found jobs on their own; five retired; and one found a job with a WREP partner.[65]

Fostering good relations with those who remained at the plants, WREP became one of the country's first private companies to sign a bargaining agreement with the American Federation of State, County and Municipal Employees (AFSCME).[66] It has rewarded staff with higher salaries and better benefits, leaving them between 9 and 28 percent ahead of their city counterparts. WREP's employees also enjoy more training and a safer work environment, bringing accidents down by 84 percent. Grievances fell from an average of 43 for the three years before privatization to an average of 0.4 for the five years following privatization.[67]

Although AFSCME formally opposes privatization – and even launched a court action to stop it – it now admits that privatization has improved the lot of its members. Union local president Steve Quick has praised both the opportunities for training and advancement and the safer work environment.[68] Union steward Cherie Moore, who once walked the picket lines to protest privatization, later embraced with the deal, urging a reporter, "don't say it's not good."[69] Former local president Stephen Fantauzzo offered the most telling comment about the workers who had moved from the city to the private firm: "The majority would say they don't want to come back."[70]

Environmental performance

Like the union activists, environmentalists initially opposed privatization. Most were concerned that a profit-driven contractor would compromise environmental safety. To ensure environmental accountability, the city provided for extensive monitoring by WREP, an independent private company, the Department of Public Works, and the Indiana Department of Environmental Management (IDEM). Two advisory panels involving academics, stakeholders, and over 20 environmental groups also monitor WREP's performance. As a result, the city has more – and better quality – data than ever and can exercise greater control.[71]

In the Fall of 1994, the death of over 513,000 fish in the White River revived controversy over privatization's effects on the environment. Frank O'Bannon, Stephen Goldsmith's successful opponent in the state gubernatorial race, tried to politicize the fish kills, blaming them on WREP's operation of one of the sewage treatment plants. The *Indianapolis Star / News* accused Mr. O'Bannon's campaign of skewing some facts and suggested that overflows from the publicly operated sewage collection system were the likely culprit. It noted that state regulators had sent a letter praising WREP for its work in meeting environmental challenges and confirming that it had found no violations of the sewage treatment plants' permits.[72] Environmental groups, including the Indiana Environmental Institute and the Hoosier Environmental Council, agree that responsibility for the fish kills rests with the city's inadequate sewer system rather than with WREP.[73] In the words of Glenn Pratt, formerly with the Environmental Protection Agency and now with Friends of the White River (FOWR), "The fish kills had nothing to do with the operation of the wastewater treatment plants. The fish kills were caused by the discharge of raw sewage from the collection system . . . [They] became very political events in which our new more political IDEM confused the facts for political mileage."[74]

The fish-kill controversy aside, privatization has clearly benefited the environment. The contract requires WREP not only to comply with all environmental laws and regulations but also to equal or better all aspects of the city's environmental performance.[75] Despite increased flows and pollutant levels, WREP has more than halved permit exceedances and has reduced fecal concentrations to a quarter of what they were under city management.[76] The plants now regularly meet the targets, established by Indianapolis, for treatment efficiency, biological oxygen demand, suspended solids, and ammonia concentrations.[77] Furthermore, they are exceeding the standards established by the Environmental Protection Agency and have earned numerous environmental awards from the Association of Metropolitan Sewerage Agencies.[78] The system is also better able to handle stormwater and has reduced plant bypasses and combined sewer overflows.

As a result, privatization has won over local environmentalists. FOWR's Mr. Pratt had vocally opposed privatization in 1993, predicting that treatment standards would be lowered.[79] He now acknowledges that WREP has improved upon the city's performance.[80] The Audubon Society's Richard van Frank agrees that operations and maintenance have improved since privatization.[81] FOWR's Brant Cowser also has praise for privatization. Citing a good working relationship and good communication with WREP, he insists that contracting out wastewater operations "was a good move for our city."[82]

The city wholeheartedly agrees. On the first anniversary of the sewage treatment plant contract, Mayor Goldsmith boasted, "we have one of the most extraordinary competition successes in the world."[83] In 1997, the city extended by 10 years the contracts to operate the collection and treatment systems. Two years later, Mayor Goldsmith expressed his ongoing delight with the arrangement: "The deal, which has proven to be a major victory for all Indianapolis taxpayers, has also benefited the local environment."[84] It is little wonder that *Public Works* magazine calls the arrangement a "model for public-private partnership."[85] It has, without doubt, been a victory for taxpayers, workers, and the environment.

CHAPTER 3

THE PROOF IN THE ENGLISH PUDDING
Debunking the Myths About Privatization in England and Wales

In 1989, Margaret Thatcher's government sold off the assets of the 10 regional water and wastewater authorities in England and Wales. That sale has become – at least in Canada – the most criticized and least understood of all privatizations. From labour and environmental activists it has received steady and harsh criticism, a good deal of which is baseless.

England and Wales privatized for many of the reasons discussed in previous chapters. A host of problems had long plagued underfunded systems. Almost a third of the treated water disappeared through leaking distribution and supply pipes, some of which dated back to the Victorian era.[1] Many sewage systems discharged untreated sewage directly into the ocean, making beaches unswimmable. The government, unwilling to insist on improvements that it would have to pay for, ignored – and in some cases, concealed – the problem.

The European Community made it impossible for Britain's government to continue denying the extent of sewage pollution. In 1975, when the EC issued a directive giving member countries 10 years to bring their "bathing waters" up to uniform standards, the government tried to evade the issue by claiming that the country had only 27 beaches. Not until 1987 did it admit that hundreds of beaches encircled its island. It then had to also admit that sewage contaminated a third of those beaches: In 1988, only 241 of 364 designated beaches met European bathing water standards.[2]

By the late 1980s, the government estimated that £24 billion would be required within 10 years to repair the water and sewage systems and to meet new European standards. However, "the financial harness of Whitehall" severely constrained any public investment. As former regulator David Kinnersley explained, "the government wanted this huge financing of additional investment to be taken out of the public sector . . . The part of it that would come from borrowing, the government wanted to be private borrowing; the part that would come from price increases, the government wanted not to be the responsibility of ministers."[3] Under privatization, the government promised, the suppliers of water and sewage services would be "released from the constraints on financing which public ownership imposes."[4]

Privatization was driven not only by mounting financial pressures but also by the growing understanding that a government could not properly regulate facilities that it owned. Indeed, Mr. Kinnersley described the regulatory regime of the time as intentionally ineffective. A "potent culture of government concealment" kept public concern at bay and enabled the government to avoid prosecuting polluting facilities.[5] In 1987, the Secretary of State acknowledged that in a publicly owned system, the government acted as both "gamekeeper" and "poacher."[6] While responsible for controlling the discharge of pollutants, it was a major discharger in its own right. These dual roles put it in an inescapable conflict of interest and made good regulation impossible. By separating the polluter from the regulator, privatization would free regulators to regulate.

To prepare for privatization and to enable the public water and wastewater authorities' private successors to meet tough environmental standards, the government wrote off £5 billion of their debts and provided them with a "green dowry" – a £1.6 billion cash injection. It then transferred their infrastructure and most of their functions to 10 new "water service companies" and sold shares in these companies in a public offering.[7] There is general agreement that the flotation price was low – to some degree, intentionally so, in order to ensure the offering's success. The shares were oversubscribed.[8]

The new companies provided sewage services to all of the connected population and water services to approximately three-quarters of the connected population. The remaining quarter continued to be served by one of 29 previously existing private water supply companies, some of which had been in business since the seventeenth century. The government established environmental, health, and economic regulators to oversee both the new water service companies and the long-established water supply companies. This combination of privatization and regulation has by many measures – including capital investment, environmental performance, drinking water quality, and customer service – been a success. By other measures – notably popularity among consumers and workers – it has fared less well.

The Benefits

A number of myths surround the benefits and costs of water and wastewater privatization in England and Wales. Anti-privatization activists routinely base their opposition to Canadian asset sales or operations contracts on England's experience, the utter failure of which has become an article of faith. Although deeply entrenched and difficult to dispel, that faith is founded on considerable misinformation.

"There is probably no more dramatic an example of a taxpayer-funded boondoggle in recent history than Britain's experience with the selling off of its water system," claimed Brian Cochrane, president of the Toronto Civic Employees Union, a decade into privatization.[9] While Mr. Cochrane's statement was more sweeping than most, his distaste for Britain's privatization is widely shared in the Canadian labour movement. The Canadian Union of Public Employees (CUPE) warned in the 2000 edition of its annual report on privatization that a "water cartel has created havoc in England" and that the UK's experience is "a clear warning sign for other countries confronting privatization."[10] According to the CUPE-led Water Watch coalition, "the disastrous consequences of privatizing water services in Britain are well known to Canadians."[11] The coalition urged municipal councils to sign an anti-privatization resolution including a "whereas" explaining that "the privatization of water delivery and wastewater treatment in Britain has led to profound problems."[12]

Union activists had an opportunity to air their concerns in 1997 when the Ontario government's Standing Committee on Resources Development held hearings into Bill 107, the Water and Sewage Services Improvement Act. The London and District Labour Council's Gil Warren told the committee that "The privatization of water failed miserably in Thatcher's Britain."[13] Rick Coronado, representing labour and environmental activists from Southwestern Ontario, reported "tragedies," warning that the "very detrimental" experience "suggests caution."[14] Sid Ryan, president of CUPE, Ontario, attested to privatization's "disastrous results."[15]

Most of the mainstream environmentalists appearing before the Standing Committee were equally critical. Both the Canadian Environmental Law Association's (CELA's) Sarah Miller and the Toronto Environmental Alliance's Janet May referred to the British privatization as "disastrous."[16] Others referred to "the horrors"of the British experience, called it "a cautionary tale for all of us," and noted it "with grave concern."[17]

Left leaning members of Ontario's provincial parliament soon took up the mantra. Marilyn Churley called the British experience "a nightmare," adding, "it went very wrong. . . . [T]here's absolutely no doubt about it. . . . [W]e all know it was a disaster."[18] Committee member Floyd Laughren, former finance minister with Ontario's NDP government, joined the chorus, calling the English privatization "a horror story" and "a debacle." He expressed surprise that so many of those who had come before the committee were familiar with the UK experience, marvelling that "An amazing number of them knew what a disaster privatization was in Great Britain."[19] The number of critics is indeed amazing, given that privatization in England and Wales has not been a disaster. On the contrary: It has brought striking economic, environmental, and public health benefits.

Canadian critics charge that the new private water and wastewater companies in England and Wales have not invested sufficient capital in the system. According to the Council of Canadians's Maude Barlow, "Little has been reinvested in the aging infrastructure."[20] CELA has likewise maintained that "Reinvestment in aging infrastructure has been meagre at best."[21] And the Ontario Public Service Employees Union (OPSEU) has claimed that "companies did little to re-invest in the water system."[22] Many of those speaking to the Ontario Standing Committee echoed such views. Representatives from both CUPE and the Canadian Auto Workers charged the water companies with failing to reinvest profits in infrastructure.[23] Toronto Councillor Peter Tabuns attributed the water companies' profitability to their reduced investment in infrastructure.[24]

Such an interpretation of the events of the last decade is startling. The 10 new water service companies have invested enormous sums in infrastructure. By 1998-99, cumulative capital expenditures amounted to £33 billion and showed no sign of letting up. That year alone, investments neared £3.7 billion – £3.2 billion for new fixed assets and £0.5 billion for infrastructure renewals.[25] By 2005, the private companies will have invested £50 billion.[26] As one official from the Department of the Environment noted, "You just couldn't contemplate that kind of expenditure in the absence of privatization."[27] Indeed, capital expenditures before privatization had been minimal in comparison, remaining well under £1 billion a year (in 1993-94 cost terms) between 1920 and 1960, and generally fluctuating between £1 billion and £2 billion a year in the 1960s, 1970s, and 1980s.[28] Annual investment averaged £1.7 billion in the 1980s.[29]

Whether the private water companies have invested their capital as efficiently as possible is subject to debate. While generally enthusiastic about the efficiencies achieved, then Director General of Water Services Ian Byatt questioned the cost effectiveness of some early expenditures, suggesting that some projects designed to improve water quality "showed poor value for money."[30] Some companies may have initially over-invested in infrastructure, thanks to incentives in their early years to "gold-plate" investment plans.[31] It is also possible that the marginal returns of more recent capital investments have declined as major environmental and health goals have already been met.[32]

Despite some inefficiencies, the water companies' investments have paid substantial health and environmental dividends. In addition to financing the construction of new primary treatment facilities for the wastewater of more than 7 million people and new secondary treatment facilities for the wastewater of more than 15 million people, the money has gone into upgrading more than 70 water treatment plants and nearly 600 waste-water treatment plants, improving more than 2,400 combined sewer overflows, and, between 1991-92 and 1997-98, building or renovating more than 46,000 kilometres of water mains and more than 10,000 kilometres of sewers.[33]

Critics of privatization in England and Wales seem unaware that the water industry's unprecedented investment and extensive program of construction and upgrades have paid off in dramatic improvements in environmental performance. Once again, the Ontario Standing Committee hearings into Bill 107 provided a forum for misinformation. The committee heard allegations of environmental deregulation, increased water pollution, numerous violations of pollution laws, environmental disaster, and other sobering effects on the environment.[34] At the end of the hearings, the committee's Mr. Laughren concluded that privatization in the UK resulted in "dirty water and unsafe sewage."[35]

The facts suggest otherwise. Environmental performance has improved by a number of measures. Between 1990 and 1998, the percentage of plants not meeting their "consent conditions" fell from 10 to less than three.[36] The total polluting loads of sewage treatment plants fell by between 30 and 40 percent during the 1990s, depending on the pollutant. Ammonia discharges fell by 37 percent. Phosphates declined by 40 percent.[37] As a result, freshwater quality improved significantly. At the time of privatization, 37 percent of the rivers and canals tested were classified as very good or good; between 1993 and 1995, that figure increased to 59 percent. The following five years saw further gains.[38] Not all waters improved: Between 1990 and 1995, the quality of about 225 kilometres of rivers and canals deteriorated. However, during that period, the quality of more than 3,000 kilometres improved significantly. In short, environmental gains outpaced losses by more than 10 to one.[39] Gains continued unabated in the following five years. In 2000, the Environment

Agency announced that rivers were the cleanest they had been since before the industrial revolution and that fish, otters, and other wildlife were returning to waters long devoid of life. The agency attribute the improvements in large part to investments made by the water companies. In Environment Minister Michael Meacher's words, "The billions being invested in cleaning up our rivers are really bearing fruit."[40]

Privatization has made coastal beaches swimmable. Between 1990 and 2000, water companies reduced the polluting load discharged to the sea by 80 percent.[41] The number of coastal beaches in England and Wales increased from 401 in 1989 to 472 in 2001. Compliance with European standards also rose dramatically, climbing from under 76 percent in 1989 to over 98 percent in 2001.[42] Thus, private operators have brought England and Wales 159 more usable beaches.

The water companies have also made considerable progress in stemming water losses. The Environment Agency's director of water management notes that a severe drought in the mid-1990s "brutally exposed industry shortcomings in this regard."[43] Spurred by public opinion and regulatory pressure, the water companies reduced leakage from reservoirs, distribution mains, and supply pipes by 9 percent in 1996-97, 12 percent in 1997-98, 11 percent in 1998-99, 7 percent in 1999-2000, and another 2 percent in 2000-01.[44] Ofwat, the economic regulator, has asked for further reductions.

Much work remains to be done before privatization can be said to have fulfilled its environmental mandate. Water companies remain among the worst polluters in England and Wales. In 2000, six of them headed up the Environment Agency's list of the 30 most prosecuted companies and another three appeared further down on the list.[45] The Marine Conservation Society complained that year that 180 million litres of raw or partially treated sewage flowed into the U.K.'s waterways or the sea each day.[46] Not until 2005 will all collected sewage be treated.[47] In the mean time, the European Commission has complained to the European Court of Justice about ongoing pollution at two beaches on England's northwest coast and has asked it to impose heavy fines on the government.[48]

However, there is little doubt that the remaining required capital improvements and repairs are more likely to be made under the new private system than under the former public system, if for no other reason than that the regulatory environment is far stricter than it was before privatization. Although bathing waters are cleaner than ever, Environment Minister Meacher has called them "not good enough" and vowed, "I want to see significant improvements and our bathing waters to be regarded among the best in Europe."[49] According to another minister, Baroness Hayman, "The government will not be satisfied until we achieve close to 100 percent compliance regularly."[50] The Environment Agency echoes such sentiments, calling for a culture change "with zero tolerance for pollution replacing apathy and acceptance of poor environmental performance."[51] Even the water companies agree that they must do better. In the words of one industry representative, "We are not happy – one incident is one too many."[52]

The rising number of prosecutions – despite the falling number of serious pollution incidents – and their higher public profile also demonstrate a tougher attitude towards pollution.[53] In 1999, the Environment Agency published its first annual "Hall of Shame" to call attention to the worst polluters, saying that "they have let down the public, the environment, and their own industry." The agency's chief executive complained that the fines imposed by the courts are too low, averaging just £3,489 for the industry and not exceeding the £36,500 paid by Wessex Water for five prosecutions: "Clearly this is not sending out a strong enough message to deter water and sewage companies who have the potential to seriously damage the environment."[54] Deterrence increased dramatically the following year, when a court imposed a record fine of £250,000 on Thames Water after a mixture of raw sewage and industrial chemicals overflowed and contaminated the Thames river and 10 homes. The Environment Agency expressed unreserved delight regarding the fine.[55]

Regulators continue to toughen their stance. In 2000, the Environment Agency announced the details of a five-year National Environment Programme to further enhance water quality. It identified all of the water companies' known environmental problems and required more than 6,000 specific projects – with an estimated price tag of £5.3 billion – to increase treatment levels, reduce storm water overflows, and otherwise clean up beaches and rivers. It expected the new requirements to bring 97 percent of the beaches in England and Wales into compliance with EC standards and to bring the water industry to the point where it would pose few threats to the water environment.[56] Indeed, the Environment Agency expects that "most of the environmental damage of the past 200 years will have been repaired by 2005."[57]

Drinking water has also improved steadily under privatization. Between

1990 and 1996, the percentage of zones fully complying with prescribed limits on individual pesticides increased from 70 to 87; on lead, from 77 to 87; on faecal coliforms, from 88 to 96; on aluminum, from 90 to 97; and on iron, from 70 to 76. Smaller improvements occurred for colour, turbidity, odour, taste, hydrogen ion, nitrate, nitrite, manganese, tri-halomethanes, and other parameters. Only in one category – PAH, or polycyclic aromatic hydrocarbons – did performance decline.[58] The Drinking Water Inspectorate reports that compliance has continued to improve, especially for pesticides – which have been virtually eliminated from drinking water – and coliforms. In 2000, 99.83 percent of the 2.7 million tests conducted met the required standards. The number of tests not meeting the standards was just one-eleventh of that in 1992.[59] The taste of water has also improved. The *Economist* reports that a test in 2000 indicated that tasters preferred London tap water to bottled mineral waters.[60] The Drinking Water Inspectorate attributes the improvements in water quality to the water companies' capital investments, to their improved planning and operations, and to its own tough enforcement actions, including prosecutions.

Within the considerable limits imposed by its near-monopoly structure, privatization has also empowered consumers. Ten customer service committees advise the economic regulator on consumer issues. Extensive consultation and information programs involve customers in both policy and performance. Dissatisfied customers have access not only to effective complaints procedures but also to redress – in the form of set compensation payments – if companies fail to meet guaranteed standards for service. These guaranteed standards require payments for interruptions in water supply, low water pressure, and flooding from sewers. They also cover customer service. For example, a company scheduling an appointment with a customer must, if requested, specify a two-hour period during which the visit will be made. The company's failure to keep the appointment entitles the customer to £20 in compensation.[61] Several companies' "customer charters" provide for compensation beyond that required by statute. Water company managers explain that satisfying customers makes good business sense, since it increases the speed with which bills are paid, reduces the costs of processing complaints, and, most importantly, opens up other business opportunities by enhancing the companies' images.[62] Both this attitude and tough regulations likely account for post-privatization service improvements, reflected in steep declines in the number of properties at risk of either low pressure or sewer flooding.[63] Whatever the cause, Ofwat's Deputy Director General suggested five years into privatization, "In many ways, better customer care has developed more significantly than any other facet of the water indus-try."[64] Increased competition for customers (as discussed in Chapter 10), will doubtless encourage water companies to provide even better service at lower cost.

The Costs

In assessing the UK's experience, business professor and sustainable development guru David Wheeler concluded, "By almost any measure, the water industry in England and Wales has achieved a great deal since privatization in 1989. However, this has not been without costs and a sig-nificant amount of conflict and political controversy."[65] Critics of priva-tization have emphasized these costs and controversies.

"When water was privatized in Britain," CUPE warned, "rates more than doubled."[66] Water Watch materials referred to annual increases of as much as 67 percent.[67] Labour and environmental activists likewise raised the alarm at the Ontario Standing Committee hearings, with one report-ing increases of 450 percent in several areas and another reporting a dou-bling of water prices in the four years following privatization.[68]

While the massive investments in infrastructure have indeed raised prices, the increases have been far more modest than critics suggest. The real average cost of unmetered domestic water services rose by 38.3 per-cent in the decade following privatization, while the real average cost of unmetered domestic sewage services rose by 46.6 percent.[69] Of course, some consumers faced above-average increases, the most dramatic of which received considerable press attention.

Many consumers – especially in the early years of adjustment – strenuous-ly objected to rate increases. Over time, however, concerns about pricing subsided. Consumers seemed to accept higher prices. Perhaps they under-stood that environmental improvements were expensive: Ninety-five per-cent of those surveyed in 1997 said that they would prefer water company profits to be spent on environmental improvements than on cuts to their water bills.[70] Perhaps they understood that costs would have risen in the absence of privatization: The real average cost of water and sewerage ser-vices to unmetered domestic customers had climbed over 22 percent dur-ing the seven years preceding privatization – and that was without major investments in infrastructure.[71] Perhaps they appreciated that private-sec-tor efficiencies offset costs that would have prompted even greater increases: Between 1993 and 1998, operating costs fell by 9 percent.[72] Director General Byatt explained that such efficiencies would greatly offset increases in bills; he predicted that 15 years after privatization, the average household bill would be £83 lower than it would have been in the absence of efficiencies.[73] Or perhaps consumers simply realized that Ofwat was determined to keep prices under control. In November 1999, the economic regulator reset price limits for all of the water companies, reducing average household bills by 12 percent in real terms and generally stabilizing them until 2005.[74] In so doing, he explained that thanks to higher efficiencies, the average annual household bill should be only £38 higher in 2005 than it was at the time of privatization.

As prices rose in the 1990s, customers were particularly sensitive to water company executives' generous salary, benefit, and bonus packages, which in some cases approached £300,000.[75] A poll commissioned by the BBC in 1998 found that while more than half of the respondents thought that they got value for money from their water companies, nearly three quarters thought that water company executives were paid too much.[76] Consumers also resented the water companies' profits and dividends. In the decade following privatization, shareholders' annual returns amounted to between 11 and 16 percent in real terms.[77] Large returns led 70 percent of those surveyed by one consumer magazine to state that shareholders have benefited more than customers from privatization.[78] Such concerns encouraged the new Labour government elected in 1997 to levy a one-time "windfall tax" on the profits of privatized water, electricity, and other utilities; the water companies' share was £1.65 billion.[79]

In the early years of privatization, the water companies' practice of disconnecting non-paying customers also drew fire. With only 8 percent of the households on water meters, most people could do nothing to keep their costs down.[80] Many had just two options: They could simply pay their rising bills or be disconnected from their water supply. Private companies had not invented the penalty of disconnection: Their public predecessors had disconnected 9,187 and 9,218 households, respectively, in the two years preceding privatization,[81] in part because changes to social assistance in 1988 had made it more difficult for those receiving state benefits to pay their water bills.[82] However, the private companies demonstrated even less patience with non-paying customers. The number of households disconnected for not paying their water bills soared to 21,282 in 1991-92. It then fell steadily, reaching 1129 in 1998-99. That year, nine of the 27 companies disconnected no households.[83] In 1999, the new Water Industry Act banned disconnection of households and vulnerable water users such as day care centres, doctors' offices, nursing homes, and schools.[84]

Canadian critics of privatization have observed that the 1992 peak in water disconnections coincided with a peak in reported rates of dysentery and hepatitis A. They have assumed that the disconnections caused the diseases and concluded that privatization itself was hazardous to Britons' health. According to CELA, "the privatization of water in England and Wales has led to profound problems such as: . . . significant increases in diseases such as dysentery and hepatitis A."[85] Presentations to Ontario's Standing Committee on Resources Development indicated that CELA was not alone. Anne Mitchell, from the Canadian Institute for Environmental Law and Policy, asserted that privatization "has led to serious public health problems." Her colleague, Mark Winfield, called privatization "a public health disaster."[86] Councillor Tabuns, who chairs the Toronto Board of Health, warned, "Given the British experience, I don't think that [the pressure to privatize] bodes very well for the health of the people who live in the Metro Toronto area."[87]

Labour unions have been particularly insistent on the ill health effects of privatization, with CUPE returning to the subject year after year. The union's Sid Ryan told the Standing Committee that privatization resulted in "outbreaks of dysentery and hepatitis A caused by poor sanitation and unavailability of water."[88] In its 1999 annual report on privatization, CUPE reported thousands of disconnections, continuing, "As a result, dysentery increased six-fold, leading the British Medical Association to condemn privatization because of the related health risks."[89] National CUPE president Judy Darcy picked up on the theme in an opinion piece in the *Halifax Herald*: "dysentery increased six-fold, causing the British Medical Association to blame Britain's private water system for increasing health risks to the population."[90] OPSEU likewise charged, "Dysentery increased by 600 percent. Hepatitis A went up 200 percent."[91]

Like the children's game of broken telephone, in which a message passed around a circle becomes increasingly garbled, the health concerns of the anti-privatization crusaders have mutated. "I think it was either typhoid or something like that which occurred," the Ontario Public Interest Research Group's Robert Barron told the Standing Committee.[92] Within six months, judging by a forum sponsored by OPSEU, the perceived health consequences of Britain's privatization had grown to include cholera.[93]

Here again, the facts do not support the alarmist claims. Anti-privatization crusaders did not imagine increases in dysentery and hepatitis A. Dysentery did indeed increase in 1992, with 16,960 cases reported that year – levels not seen since the 1960s, when between 19,491 and 43,285 cases were reported annually. Hepatitis A also increased in 1992, by 426 cases, to 7,856. However, privatization's critics do not mention that in 1993 hepatitis A cases fell to 4,457 – just 84 percent of the 1989 pre-privatization baseline. They fell further in the following five years, staying at between 25 and 51 percent of that baseline. In 1996, 1997, and 1998, dysentery rates were also lower than they were at the time of privatization.[94]

Researchers were unable to find any causal relationship between disconnections and disease. In 1993, the British Medical Association asked its Board of Science and Education to prepare a report on the health consequences of water disconnections. While the report stressed water's vital role in preventing disease and recommended making disconnections illegal, it concluded that "a causal link has yet to be established between water disconnections and infectious diseases, such as dysentery and hepatitis A." It noted that increased rates of the two diseases could be considered part of long-established cyclical patterns. It also noted that dysentery had increased in Scotland, where water suppliers do not disconnect non-paying customers.[95]

Other researchers reached the same conclusions. In Sandwell, where high numbers of both water disconnections and cases of dysentery and hepatitis A appeared within the same post codes, doctors reported that none of the reported infections occurred in a household where the water had been cut off. Although the Director of Public Health wondered if other cases may have gone unreported, he could conclude on the basis of the evidence only that "there is a direct association between both diseases and poverty. Families who are poor and more likely to be at risk of having their water cut off, are also at most risk of these infections." Britain's Chief Medical Officer also reviewed the possible relation between gastrointestinal infections and water disconnections, finding in 1992 that "there is no evidence at this time stage that the two are connected." The following year, the Faculty of Public Health Medicine, while stressing its opposition to disconnections, likewise stated, "we accept that a convincing case has not yet been made on epidemiological grounds."[96] The economic regulator confirmed such conclusions in 1999: "Ofwat has seen little evidence of a link between water disconnections and public health. . . . [T]he issue is not mentioned in any Ofwat documents on disconnection."[97]

Although anti-privatization crusaders have bungled the facts about investment and performance, they have been right about one issue: Privatization has reduced the number of jobs at the water utilities. In 1989, the water service companies employed 47,807 people. By 1998, that number was down to 31,310, a reduction of almost 35 percent.[98] The companies continue to cut staff: In December 1999, five companies responded to planned reductions in water prices with announcements of 3,200 layoffs. Analysts assume that further layoffs will follow.[99]

Organized labour admits that various benefits have helped offset the job losses. While confirming its opposition to privatization, the Amalgamated Engineering and Electrical Union notes that severance and early retirement packages have often been generous, making job losses tolerable if not welcome. It also notes that remaining employees have gained from their shares in the privatized companies and that, in some companies, the employees are better paid, better trained, and enjoy better working conditions.[100] The Reason Foundation points out that workers outside of the water utilities have also benefited. Workers in the construction industry have gained from the upgrading of the water and wastewater systems. And those in export or consultancy have gained from the British companies' new international prominence.[101] The creation of jobs elsewhere, however, is cold comfort to those within the industry. CUPE's Sid Ryan spoke to the Ontario Standing Committee on Resources Development not of the jobs created but of the thousands of layoffs. He frankly admitted his union's concern that the same might happen here in Canada: "As workers, we fear the loss of our jobs as a result of privatization."[102]

If the water companies' role is to keep as many people as possible employed, privatization in England and Wales has failed. If, on the contrary, their role is to bring capital to a system long starved of cash, to upgrade and repair crumbling infrastructure, to clean up rivers and beaches, and to provide better water and better service to their customers, privatization looks much more like a success. As the *Economist* concluded in 2001, "For consumers, water privatization has, by and large, worked."[103] Neither the water companies nor the environmental or economic regulators have yet achieved all that they set out to achieve. As the environment department's Michael Williamson explained, "we're still in the early days of feeling our way. Let's face it, this was the most radical and most complex privatization that's ever been adopted in the world and I don't profess that we've got it absolutely right." However, his caution hardly tempered his enthusiasm: "we've got such a wonderful water industry in England and Wales that I can go on madly about privatization."[104] Reviewing the first decade of private ownership and operations, Ian Byatt, then the water industry's chief economic regulator, was only slightly more restrained. The privatization of the water companies, he said, coupled with their regulation under a system that acts at arm's length from government, had resulted in dramatically improved service delivery, much greater efficiency, steady prices, and good returns to shareholders. He concluded, "There have been spectacular successes."[105]

CHAPTER 4

EAU NO
The Deficiencies of the French Model of Water and Wastewater Privatization

In stark contrast to England and Wales, France has taken a decentralized and evolutionary approach to privatization. The country's 36,500 communes, or municipalities, make their own decisions about who should build and operate their publicly owned water and wastewater systems.

Like the United States and England, France has a long history of privately supplied water. Two giants of the French water industry – Compagnie Générale des Eaux (now part of Vivendi) and Lyonnaise des Eaux (now Suez) – entered the business in 1853 and 1880 respectively.[1] The French water companies steadily increased their participation in the market over the years. Business boomed after World War II, since the war had not only destroyed much infrastructure but had also weakened municipalities' ability to rebuild. By 1954, private companies provided 31 percent of the country's water supply; the figure had increased to 60 percent by 1980.[2] Another boom occurred in the 1980s and 1990s in response to municipalities' needs to meet financially and technically demanding environmental directives.[3] Today, private water companies, operating municipally owned facilities under a wide variety of short- and long-term contracts, serve over 79 percent of France's population, while private sewer companies serve 40 percent.[4] Generally, only rural municipalities continue to manage water services directly.[5]

Although the sheer numbers make it impossible to identify a specific set of reasons that apply to all privatization decisions, economic and environmental factors are generally similar to those motivating privatization elsewhere. Over the last 25 years, France's Cour des Comptes – similar to Canada's Auditor General or the U.S. General Accounting Office – has periodically called attention to the shortcomings of public provision of water and sewer services, documenting extensive failures to provide safe drinking water and adequate sewage collection and treatment.[6] Public utilities, over-staffed with insufficiently knowledgeable and unmotivated employees, have been inefficient and ineffective in constructing and operating systems. In 1980, only 69 percent of France's coastal beaches met European bathing water standards, in large part because they were contaminated with raw sewage.[7] By 1995, although compliance had improved considerably, over 7 million people continued to discharge their wastes, untreated, into local waters.[8] Problems likewise beset drinking water systems, which sporadically provided substandard water to more than half of their customers.[9] Leaks from ill-maintained nineteenth-century water supply pipes, along with poor metering, accounting, and collection practices, have resulted in high levels of unaccounted-for water.[10]

Meanwhile France, like England and Wales, has faced demanding environmental and health directives issued by the European Union. Directives governing drinking water, bathing waters, surface waters, fish and shellfish waters, sewers, and sewage sludge have strained many communes' financial and technical resources. Particularly challenging has been the 1991 Urban Wastewater Treatment Directive (amended in 1998), requiring secondary – or, in sensitive zones, tertiary – treatment for the wastewater of inland communities with over 2,000 inhabitants. When the directive was transposed into French law in 1992, only 57 percent of those affected by it were served by treatment plants deemed to be in compliance. Facing deadlines for compliance ranging from 1998 to 2005, communes could no longer avoid building new treatment plants.[11] Eager for the technical expertise of water industry specialists, wanting to benefit from the economies of scale and other efficiencies enjoyed by the large water companies, and in need of private funds to supplement water revenues and public subsidies, they have increasingly turned to the private sector for assistance.[12]

Privatization in France appears to have been somewhat less successful than that in England and Wales. Despite considerable private sector involvement, the French water and wastewater systems remain surprisingly public. All of the assets, be they publicly or privately operated, remain under communal ownership. Subsidized and politicized, the systems are largely insulated from market forces.

The privatization of the delivery of French water and wastewater services has not led to the privatization of the financing of these services. The systems are heavily subsidized, with public and private operations alike receiving assistance from several levels of government and from one another. In 1991, the water companies themselves funded only one-third of that year's capital expenditures on water and sewage.[13] Water consumers regularly subsidize other water and wastewater services consumers within their river basin. Water companies pay withdrawal and pollution fees to one of the six Water Agencies, the country's river basin management authorities. The agencies then reinvest the fees in water and sewage projects. The fees, which amounted to FF9.4 billion in 1995,[14] support both grants and low-interest loans, either of which may cover up to 50 percent of the capital cost of a project.[15] Subsidized water projects include dams, reservoirs, interconnections between systems, water intakes, and demand management programs; subsidized sewage projects include sewer systems, treatment plants, and outfalls. Between 1997 and 2001, the agencies contributed FF54 billion towards sanitation.[16]

All levels of government also provide subsidies. Communes contribute at least 20 percent of construction costs.[17] Departments, which distribute national subsidies, provide billions each year for water and sewage investments; many have also established Technical Assistance Services for sewage treatment. In 1996, departments spent FF4.7 billion on water and wastewater management and millions more on surface and groundwater protection.[18] Regions occasionally help fund some of the larger projects.[19] In 1996, regions spent more than FF213 million on water and wastewater management and the same amount on surface and groundwater protection.[20] The national government subsidizes water projects in rural communities through *le fonds national du développement des adductions d'eau*, the national fund for the development of water supply systems. It collects money from water users through a surtax on water bills; it also collects money through the national tote betting system.[21] This dizzying array of subsidies has left consumers bearing only a small fraction of water and wastewater costs: In 1999, households paid just 8 percent of the latter.[22]

Despite generous subsidies, prices for both public and private services have risen dramatically in recent years, as providers have worked to comply with European Union standards.[23] Although Jihad Elnaboulsi, an economist with the French government, reports that many studies have shown that delegation of water and wastewater services to the private sector brings higher efficiencies and more benefits,[24] there is no evidence that private sector efficiencies have reduced prices for French consumers. In fact, the private water companies have generally charged higher prices than the public operations. In 1991, the price charged by private operators was 23 percent higher than that charged by public operators. In 1996, the difference had fallen to 16 percent.[25] A number of factors may help explain these differences in costs. Higher prices in the private sector may reflect political costs that are unrelated to the actual provision of water and wastewater services. In the past, private companies frequently paid entry fees to communes; they subsequently recovered these payments – at times amounting to hundreds of millions of francs and going towards municipal budgets rather than water works – through elevated water prices. In some cases, political contributions and even bribes have further increased the costs of doing business.[26] To the extent that such expenditures have been responsible for higher prices, recent legislation prohibiting entry fees[27] and limiting political donations[28] should reduce prices accordingly.

The private sector's higher prices may also point to the limited nature of the competition within the French water sector. Three well-established companies hold over 96 percent of the contracts for water supply.[29] France's Cour des Comptes has complained of a perennialization of the arrangements between municipalities and private providers of water services. The court's 1997 report noted that competition is often "organised" and that negotiations may take the place of competitive bidding, resulting, at times, in the entrenchment of particular companies.[30] One company has managed water services for Ile-de-France's 4 million customers for over 47 years without competition. The company serving Dinard has, without competition, provided water since 1929 and will continue to do so until at least 2005.[31]

Alternatively, the higher rates charged by private operators may be explained by their providing a higher level of service. Communes may operate their systems when doing so is easy and inexpensive; they may turn to the private sector when they encounter technical difficulties or are required to upgrade – in short, when operations become more costly. Differences in accounting practices may also help explain the differences in prices. If municipalities do not count all of their costs, such as depreciation, they will appear to have fewer costs to pass on to consumers.[32] These factors make it difficult to draw conclusions about relative efficiencies obtained by the private sector.

Government subsidies and rising consumer prices have not produced service of the caliber that their levels might suggest. Utilities commonly provide drinking water that fails to meet health standards. In 1995, the latest year for which figures are available, more than 24 million people – 55 percent of the connected population – at least occasionally received substandard water. Almost 16.7 million people received water that sporadically failed to meet bacteriological standards; almost another million received water that failed to meet such standards for more than 30 days. The water provided to more than 8.5 million people exceeded turbidity limits, while that provided to almost 5.4 million exceeded pesticide limits.[33] As unimpressive as these figures are, those from previous years were even worse. Between 1993 and 1995, approximately 30 million consumers connected to systems serving more than 5,000 people received substandard water.[34] Unsurprisingly, many of the French have given up on tap water. An environment institute survey conducted in 2000 found that just 58 percent of the respondents normally drank tap water and just 31 percent did so exclusively. While most of the abstainers disliked the taste of tap water, many also cited concerns about the water's calcareousness and the health risks it posed.[35]

Progress is more evident in the country's wastewater systems, in which tremendous investments have been made in recent years, primarily to enable plants to comply with European requirements. These investments are paying off in cleaner beaches. France boasts that, in 1998, 94.5 percent of its coastal beaches – up from 93 percent the previous year – complied with European standards.[36] According to the European Commission, compliance was somewhat lower, at 94.5 percent and 90 percent respectively.[37] Measurement against European standards can be difficult. France analyses faecal streptococci and faecal coliforms, but not total coliforms, takes fewer samples than recommended by the commission, and calculates compliance levels differently than the commission, taking into account only selected sampling points rather than all points.[38] Since its figures do not include permanently closed beaches, they certainly do not reflect the full impact of sewage pollution on coastal communities. They do, however, suggest progress. Furthermore, given that, in 1999, remedial measures had been implemented in almost 83 percent of the surveyed areas not complying with standards, greater progress can be expected. Because industrial action by staff prevented France from submitting to the European Commission data for either 1999 or 2000, it is not yet possible to determine whether those expectations have been met.[39]

Another sign of progress appears in the number of sewage treatment plants being equipped to remove nitrogen and phosphorus. In 1986, 62 plants had the capacity to treat nitrogen, and just 22 plants could treat phosphorus. By 1995, those numbers had increased to approximately 150 and 200, respectively.[40]

Nonetheless, inadequate sewage treatment remains a problem. In 1996, in urban areas of more than 10,000 people, 33 percent of the flow of oxidisable material – raw organic sewage from domestic sources and from industrial sources connected to municipal sewers – was not collected by treatment plants. This pollution – including waste from collection networks not connected to treatment plants and overflows of combined sanitary and storm sewers – was returned directly to the environment without treatment. Another 18 percent of the oxidisable material reached the plants but was not removed. Either the plants were not big enough to handle flows, or they used old, unreliable equipment. Whatever the reason, treatment plants eliminated only 49 percent of the organic sewage produced, while 51 percent – equivalent to the waste from almost 29 million people – was released into the environment.[41]

Le Réseau National des Données sur l'Eau, France's national network for water data, publishes records on wastewater collection and treatment for communities of over 10,000 people. Although it also publishes information on which, if any, private companies serve each community of over 40,000, it does not correlate the data. It is clear, however, that inadequate treatment plagues both public and private operations. While the publicly managed facilities in Vichy, Lisieux, and Calais removed just 30, 36, and 39 percent of oxidisable material, respectively, in 1995, some of the privately managed facilities fared even worse: Concarneau, Vernon, Nantes, Rouen, Dieppe, and Pau removed between just 10 and 28 percent.[42]

Those responsible for such sewage pollution are rarely prosecuted. The Organisation for Economic Co-operation and Development (OECD) noted in 1997 that authorities are reluctant to sanction municipalities. The country's extensive array of laws forbidding pollution are rarely enforced against municipalities; "genuinely dissuasive sanctions are seldom used."[43] The OECD attributed local problems to a lack of both funding and political will. In fact, the two are linked, given the public ownership of all sewage facilities in France and their wide-spread reliance on public subsidies. In a report to the United Nations, Mr. Elnaboulsi acknowledged the potential problems inherent in public ownership, noting that municipalities that both own and operate their treatment systems run "the risk of a potential conflict of interest."[44] He noted that a 1995 law tried to remedy the problem of municipal pollution with provisions for the prosecution of local executives and mayors. However, once the government saw how effective the law could be – several mayors were prosecuted for sewage pollution – it backtracked. Indeed, in 1996, it formalized its protection of some polluting municipalities, passing legislation limiting officials' criminal responsibility for harm wrought by their non-negligent actions. The Minister of the Interior assured a gathering of high-ranking civil servants that the new law was immediately applied to cases in progress and made it possible to exonerate some of their colleagues.[45] Public prosecutors are now advised to take account of mayors' human and financial resources and other inherent difficulties in their missions before assigning responsibility for pollution.[46]

The politicization of France's water and wastewater systems has contributed to their failure to become either financially independent or environmentally sound. To the extent that politics drives water subsidies and permits sewage pollution – to the extent, in other words, that the problems associated with France's water system reflect government failure – only full privatization, or the complete withdrawal of government involvement from the finance and operation of the system, will correct the problems.

PART II

WHERE'S CANADA?

CHAPTER 5

A GROWING CRISIS
The Need for Private-Sector Involvement in
Canada's Water and Wastewater Systems

"Canadians have been slow to capitalize on the benefits of public-private partnerships in water and wastewater." So lamented Thompson Gow and Associates in a study prepared in 1995 for Environment Canada. The authors warned, "If present trends continue, Canadians may soon be facing a growing crisis in water and wastewater infrastructure. Aging infrastructure, increasing demand for services, and reduced government resources are all major contributing factors to this crisis. As experience in France, England and Wales, and the United States demonstrates, private sector participation in the water and wastewater industry could be an important part of the solution to the problems facing Canada."[1]

The crisis that Thompson Gow and Associates warned of is upon us. The Walkerton tragedy called attention to severe problems plaguing water infrastructure all across Canada. In the months following the tragedy, 246 boil-water advisories were issued in Ontario, most often in response to high total coliform counts.[2] Inspections revealed deficiencies in sampling, maintenance, training, or performance at 357 of Ontario's 645 water treatment plants.[3] The Quebec government urged consumers served by 90 systems in need of urgent repairs to boil their drinking water,[4] while the government of Newfoundland decided that it would be "prudent and responsible" to issue boil-water advisories for 188 communities with inadequate or no chlorination.[5] Similar advisories covered more than 200 communities in B.C.[6] Meanwhile, a report warned the Saskatchewan government that 121 rural communities had deficient water treatment systems.[7]

Problems beset Canadian drinking water systems long before Walkerton brought them to the forefront of public attention. In 1985, the Federation of Canadian Municipalities estimated that some or much repair was needed on at least 31 percent of Canada's water distribution infrastructure, which then averaged 30 years of age.[8] (Water supply pipes continue to leak badly. On average, 14 percent of municipal piped water is lost in leaks; in some communities, such as Montreal, the figure is as high as 40 percent.[9]) Even well maintained pipes and plants have proven inadequate to meet the needs of growing urban populations. Many municipalities have lacked the infrastructure required to treat, store, and transport drinking water. In 1994, 17 percent of Canada's municipalities with water systems reported problems with water availability.[10]

The shortcomings of Ontario's drinking water systems have been known for a decade. A 1992 report by the Ministry of Environment and Energy revealed that fewer than half of the water plants assessed complied with provincial policies regarding surface water and groundwater or with the health related parameters found in the Ontario Drinking Water Objectives.[11] The health risks of inadequate or poorly performing systems soon became apparent. In 1993, cryptosporidium sickened members of almost 24,000 households in Kitchener-Waterloo.[12] (That year demonstrated the potentially lethal consequences of such a contamination, when the parasite sickened 406,000 residents of Milwaukee, Wisconsin, sent over 4,000 to hospital, and killed 111.[13]) After the 1996 cryptosporidium outbreak in Collingwood, Ontario, the provincial government admitted that 43 of the plants it operated were vulnerable to the parasite.[14] A variety of contaminants, ranging from merely unpleasant to deadly, have plagued other systems. A report compiled in June 2000 by Ontario's environment ministry chronicled 108 "adverse water incidents" across the province. The report described the contamination of systems in communities, trailer parks, subdivisions, schools, and seniors' residences by E. coli, coliform, cryptosporidium, giardia, arsenic, gasoline, and PAH.[15] The solvent trichloroethylene, a suspected carcinogen, has also been found in the water of a number of communities.[16] One explanation of the province's wide-spread problems may lie in the fact that many of the people operating drinking water plants lack formal education and training. In the early 1990s, when the province introduced licensing of plant operators, it grandfathered 2,081 workers, certifying them if they had as little as one year's experience on the job.[17]

Quebec's drinking water systems have likewise been inadeqaute for some time. Between 1989 and 1995, 45 water-borne epidemics, sickening 1,800 people, were reported in the province.[18] Water related illnesses are likely more widespread than such figures suggest. After studying 45 water treatment plants along the Saint Lawrence River, microbiologist Pierre Payment estimated that drinking water meeting current guidelines could be responsible for one-third to one-half of the gastrointestinal illnesses of local populations.[19]

Other provinces and territories have faced a wide array of problems with their systems. A 1995 federal government study found that the water systems in 171 aboriginal communities – or one in five – posed a health risk.[20] In 1996, the threat posed by cryptosporidium prompted environmental toxicologist Joseph Cummins to tell the *Globe and Mail*, "We've recently realized that Canada's water-treatment facilities are nowhere near adequate."[21] That year saw the infection of up to 10,000 Kelowna residents by cryptosporidium and the infection of 3,000 southern Vancouver Island residents by the toxoplasmosis parasite.[22] Boil-water advisories became commonplace in the province. An average of 224 communities were under advisory at any one time between 1994 and 2000; the number increased to 304 in 2001. In his Annual Report for 2000, B.C.'s Health Officer noted that there had been 29 confirmed waterborne disease outbreaks in the province since 1980 and that the province had long had the highest rate of intestinal illness in the country. His blunt assessment: "In many areas of the province, B.C. has been under-treating its water for years."[23] On the other side of the country, in December 1999, random testing revealed unacceptably high levels of trihalomethanes – chlorination by-products that may cause bladder cancer – in 18 Nova Scotia reservoirs.[24] The following month, the Newfoundland government announced that the tap water in 63 communities had dangerously high levels of trihalomethanes. This followed an earlier report condemning the drinking water in 120 Newfoundland communities, some of which have had to boil their water for the last 10 years.[25]

Sewage pollution is also a serious problem across Canada. Approximately 75 percent of Canadians are connected to sewer systems. Not all of those sewer systems, however, are connected to treatment plants. Seven percent of them simply carry the sewage, untreated, to the nearest lake, river, or ocean. Over one-and-a-half million Canadians dispose of their sewage in this way.[26] Only 67 percent of urban Canadians living east of the Ottawa River are served by sewage treatment plants.[27] Every day, 194 Atlantic communities pump raw sewage into the ocean.[28] The region's larger cities are among the worst offenders: Saint John, New Brunswick, provides sewage treatment for less than half its residents; Halifax and Dartmouth, Nova Scotia, provide no treatment for central city residents; and St. John's, Newfoundland, provides no treatment at all. The Sierra Legal Defence Fund's 1999 report card on the treatment methods and discharges of Canadian cities paints a grim picture: "Of the 21 cities documented in this report, five (Victoria, Saint John, Halifax, St. John's and Dawson City) dump a combined total of 365 million litres of untreated sewage directly into the nation's rivers, lakes and seas every day. Eleven other cities dump an average of 437 million litres of untreated sewage per day through by-passes and combined sewer overflows."[29]

Connection to a sewage treatment plant does not guarantee adequate treatment. Five million Canadians – almost a quarter of our municipal population – have access to only primary treatment.[30] A conventional primary plant removes only 40 to 60 percent of the suspended solids and only half of the total coliforms; it reduces biological oxygen demand by only 25 to 40 percent.[31] According to the Sierra Legal Defence Fund, Montreal, Charlottetown, and Vancouver discharge 2.4 billion litres of primary effluent daily.[32]

Even the more common secondary plants – or the sewers serving them – may be too small, poorly constructed, ill maintained, or badly operated. This poor state of repair may help explain why so many plants across the country fail to comply with provincial standards. In some provinces, such as Nova Scotia, non-compliance is the norm.[33] Others provinces, such as Quebec, fail to adequately monitor or report on plant performance.[34] Even the provinces that monitor their plants tolerate ongoing violations. In 1998, 56 of Ontario's sewage treatment plants failed to comply with legal requirements, policies, or guidelines. Many of the non-compliant municipalities were repeat offenders: 25 had appeared on the 1997 list, and 35 had offended at least once during the previous six years.[35] Chronic offenders have likewise graced British Columbia's non-compliance list. One of Vancouver's plants appeared on 14 lists before finally making the grade.[36] The provincial government does not appear to be in any hurry to fix Vancouver's sewage system: In 2002, it approved a plan that gave the region 18 years to clean up the discharges from one plant, 28 years to improve another, and 50 years to stanch combined sewer overflows.[37]

In 1999, the failure of 19 municipal sewage operations to meet British Columbia's standards prompted Environment Minister Joan Sawicki to express particular concern about the potential environmental impacts of sewage discharges.[38] Ms. Sawicki is right to worry. Untreated or inadequately treated sewage contaminates water supplies. It harms fish and fish habitat. It sickens swimmers and causes frequent beach closures. Shellfish closures are also common. On the Atlantic coast, it is not unusual for thousands of square kilometres of shellfish grounds – more than a third of the region's total – to be closed. Municipal sewage pollution is the sole cause of 20 percent of these closures and contributes to over 50 percent of them.[39] Municipal sewage pollution has likewise closed 18 percent of Quebec's soft clam and blue mussel harvesting zones.[40]

Fixing Canada's water and wastewater problems will require billions of dollars. The National Round Table on the Environment and the Economy suggested in 1996 that over the following 20 years, Canada would need to invest between $38 billion and $49 billion to maintain and refurbish existing water and sewage infrastructure. In addition, it estimated, it would need to invest $41 billion in new stock. These investments would be required to meet existing standards; tighter standards would require even higher investments. The Round Table stressed that its estimates were conservative, noting high-end projections for new infrastructure require-ments of $100 billion.[41] The Canadian Water and Wastewater Association roughly echoed the Round Table's projections, estimating that between 1997 and 2012, $27.6 billion would be required to renew water treatment and distribution and $61.4 billion would be needed to upgrade sewers and wastewater treatment. It warned that further investment would be required to serve an expanding population or to meet more stringent standards.[42] Such estimates – based on almost no detailed information on the condition of the country's water and wastewater infrastructure – are not reliable. Nor do they account for the potential of accounting and pricing reforms to obviate the need for system expansion. Regardless, it is clear that very substantial investments are required.

The tens of billions required are not, however, readily available. Traditionally, municipal water and sewage systems have received gener-ous financial assistance from higher levels of government. Two decades ago, federal assistance comprised more than half of total government funding for water and sewage treatment.[43] Federal assistance declined in the 1980s, while provincial subsidies gained importance. In the 1980s and 1990s, it was not unusual for Ontario to pay for 85 percent of a sys-tem's capital costs; in some cases, Ontario grants comprised 92.5 percent of total capital costs.[44] In the mid-1990s, Quebec financed between 85 and 94 percent of construction costs, Newfoundland financed between 30 and 80 percent, New Brunswick financed up to 80 percent, British Columbia financed between 25 and 75 percent, and Nova Scotia financed 50 percent.[45] The National Round Table estimates that, in 1994, con-sumers benefited from $5.3 billion in operational and capital subsidies to water and sewage projects.[46]

Generous subsidies turned a municipal councillor's main job, in the words of Ontario MPP Doug Galt, into lobbying for money from the provincial government.[47] Sam Morra, executive director of the Ontario Sewer and Watermain Construction Association, described the conse-quences: "Smaller communities have become addicted to the provincial subsidies."[48] The addiction remains much in evidence. Michael Power, president of the Association of Municipalities of Ontario, charged in June 2000 that "the single most important impediment to the successful maintenance and rehabilitation of Ontario's municipal infrastructure is a shortage of funds." Insisting that the problem cannot be fixed by individual municipalities, he urged the immediate establishment of a tripartite infrastructure program involving the federal, provincial, and municipal governments. He pleaded, "We are standing by the phones, Mr. Premier. We are waiting for your call, Mr. Prime Minister."[49]

Mr. Power may be waiting by the phone in vain. In recent years, subsidies have declined. They have not completely disappeared. The federal government promised a new Strategic Infrastructure Foundation in its 2001 budget.[50] A portion of the minimum of $2 billion to be spent will undoubtedly subsidize water and sewage projects. Provincial subsidies have also continued to trickle out, as was evidenced by Ontario's 1997 announcement of a three-year, $200-million Water Protection Fund to help municipalities upgrade their water and sewage treatment systems and British Columbia's 1998 announcement of a three-year, $150-million grant program for water and sewage projects.[51] But even after the Walkerton tragedy focussed attention on the abysmal state of water systems, Ontario was willing to commit just $240 million, through the provincial SuperBuild Corporation, to health and safety infrastructure – a small fraction of the tens of billions needed to repair the province's water and wastewater infrastructure.[52]

As municipalities struggle to meet their needs with less help from upper levels of government, their other traditional financing sources – taxes and debt – are also increasingly constrained. The downloading of responsibilities from provincial governments has stressed municipal budgets and rising tax rates have led to "ratepayer fatigue."[53] Politicians are also reluctant to tap consumers for their share of the costs. Consumers have long been spoiled by some of the lowest water prices in the industrialized world – prices that have not even begun to cover the full costs of operating their systems, let alone the costs of upgrading them.[54] In the mid-1980s, the Federation of Canadian Municipalities estimated that user charges covered just 50 percent of sewage collection and 65 percent of wastewater treatment operations.[55] Based on 1991 data, economist Steven Renzetti determined that residential and commercial prices in Ontario were only one-third and one-sixth of the estimated marginal cost for water supply and sewage treatment, respectively.[56] Such low prices give Canadians little incentive to conserve, prompting water use rates that are the second highest in the world and twice the European average.[57] The

OECD reports that underpricing of water in Ontario has caused consumption to exceed efficient levels by an average of 50 percent.[58] Of course, overuse places higher demands on systems, further increasing the need for investment in infrastructure. The Canadian Water and Wastewater Association estimates that average household bills would have to rise from $0.90 per day to $2.46 per day to meet investment needs.[59] Yet politicians resist raising rates, fearing a voter backlash and fearing that industrial and commercial users will vote with their feet.[60] As the Quebec government's Commission on Water Management notes, engaging in costly infrastructure repair is "politically speaking, unprofitable" for municipalities.[61]

It is becoming increasing clear, the Thompson Gow study prepared for Environment Canada concluded, that "the public sector cannot meet the required levels of capital investment in water and wastewater infrastructure." Within government bureaucracies and consulting circles, there is widespread recognition that privatization could help solve this and the other problems troubling our water and wastewater systems. Throughout the 1990s, numerous studies of privatization resulted in almost as many endorsements of the process. The studies identified a tremendous range of potential benefits. Thompson Gow and Associates enumerated the economic and environmental benefits as follows: higher levels of financing; shorter construction schedules; greater incentives to implement new technologies; reduced maintenance costs; increased tax revenues; revenue windfalls from asset sales; better valuation of water resources; and improved environmental performance. It also noted that public-private partnerships in Canada would improve the domestic water industry's capacity to export services to growing world markets.[62]

The National Round Table on the Environment and the Economy stressed this last point in its Sustainable Cities Initiative, where it promoted the private provision of water and wastewater systems in Canada as a way of positioning Canadian firms abroad to meet the growing demand for private involvement in water and wastewater infrastructure. Before they can compete to provide services abroad, Canadian firms need experience at home. "The most important action that Canadian governments at every level can take is to use the PPI [public-private infrastructure] model themselves," the Round Table explained, adding, "Canada has a window of opportunity to position itself as a front-runner, rather than an also-ran, in providing real urban solutions. We just need to get started – now."[63] The interest in opening up a major export market also appeared in the Round Table's 1996 report on water and wastewater services. Perhaps more important was that report's acknowledgement of the domestic gains to be had from privatization. While noting the lack of a national consensus on privatization, the report concluded, "given public fiscal realities, a major infusion of private capital is required to maintain existing systems and build new facilities."[64]

Among the government agencies persuaded of privatization's advantages was Industry Canada, which saw in it potential for meeting domestic environmental needs while building a base for the export of environmental services. In 1995, the agency sponsored a series of workshops across the country to stimulate interest in public-private partnerships for municipal environmental infrastructure and services. The Vancouver workshop helped prompt a federal-provincial initiative to involve the private sector in solving British Columbia's wastewater treatment problems. The report of a subsequent "awareness workshop" touted public-private partnerships – covering a broad continuum from operations and maintenance contracts to build-own-operate agreements – as a "robust and flexible" framework for dealing with municipal sanitation. Their benefits, the report elaborated, include: access to capital; enhanced debt ratings; otherwise unaffordable investment in new or improved facilities; more rapid development; more efficient operation; greater cost control; new revenues; reduced public-sector risk; improved valuation and accounting of water resources; improved environmental performance; increased quality control; better asset preservation; deep technical expertise; greater incentives for technological developments; and improved capacity for domestic companies to compete internationally.[65]

On the other side of the country, municipalities were being bombarded with similar information. In 1994, Canada and Nova Scotia joined together to promote private sector participation in municipal infrastructure. The resulting three-year, $4-million cooperation agreement helped fund 37 pilot projects – including three in the water and wastewater sector – and several resource documents to help municipalities negotiate public-private partnerships.[66] One of the guides produced under the agreement described the by now familiar array of benefits from privatization, including: construction cost savings; operational savings; faster implementation; preserved or improved levels of service; risk sharing; wide range of financing options; avoidance of capital debt; enhanced public management; greater performance measurement; increased public sector revenues; enhanced economic development; innovative solutions; enhanced facility maintenance; true costing of services; and arms-length independence, or the de-politicization of service delivery.[67]

In the early 1990s, Ontario's Ministry of Municipal Affairs heard the same message from Price Waterhouse after commissioning a study of innovative financing approaches for municipal infrastructure, including water and wastewater systems. The consultant sang the praises of voluntary private sector participation, calling it "perhaps the most desirable mechanism for funding municipal infrastructure." Among the significant benefits for the public sector were: access to capital; access to technology; revenue enhancement; risk allocation; increased efficiency; and reduced construction costs and times.[68]

The message resurfaced in the report of the Provincial-Municipal Investment Planning and Financing Mechanism Working Group, whose membership included representatives of local and regional municipalities, school boards, and the Ontario government. The report noted that the private sector is often better positioned than the government is to manage the risks associated with project financing, operating, marketing, and regulation. Communities taking advantage of the private sector's experience and skills, flexibility, and access to funding could look forward to undertaking projects that could not otherwise have gone forward, to lower project and operating costs, and to re-directing government resources to other pursuits. Furthermore, firms could develop highly exportable expertise.[69]

Throughout the 1990s, the Ontario government was inundated with briefings, reports, and studies on privatization. The MISA Advisory Committee, the Interministry Committee on Local Government, the Alternative Financing and Public-Private Partnership Working Group, the Secretary of the Management Board, and numerous consultants advocated private financing and competitive contracting, promising greater expertise, access to new technologies, incentives to innovate, increased efficiencies, faster delivery, lower operating costs, improved service, and economies of scale.[70]

Despite a seemingly endless stream of studies endorsing the concept, however, privatization remains the exception to the rule in Ontario, and, more generally, throughout Canada. This is particularly puzzling in Ontario, where the government formally embraced privatization in 1996. That year, in a confidential "minute," the provincial Cabinet agreed that "opportunities for private sector participation in financing and delivering water and sewage services should be enhanced."[71] A briefing note of that time referred to increased private sector involvement as "a key provincial objective."[72] The privatization policy was quickly embraced by the

Ministry of Environment, which vowed to "encourage, facilitate and support new approaches and partnerships."[73] The policy of increasing private sector involvement retained its importance – in planning documents if not in reality – in the following years. "Enhancing private sector participation and competition" is listed as a policy objective in a May 1997 submission to the Cabinet Committee on Privatization.[74] In or after 1999, private investment was to be included in the government's longer-term strategy: "The Ministry, along with SuperBuild, the Ministry of Finance, MMAH and others plan to develop a longer term strategy . . . Such a strategy would focus on a number of principles and criteria, including: . . . encouraging private sector involvement in investing in water and sewage infrastructure."[75]

Although enhancing private sector involvement was clearly government policy, the government failed to communicate the policy to municipalities or the public. With the exception of the tentative and ultimately ineffective initiatives discussed below, it took no action to implement the policy. When asked at the Walkerton Inquiry if enhancing opportunities for private sector participation remained a goal during his tenure, former environment minister Norman Sterling replied, "Well, nothing happened really in that area." When pressed, "was there any encouragement of any private sector involvement?" he replied, "Not that I'm aware of."[76]

In 1997, the Ontario government passed a law that many thought would hasten the privatization process. Bill 107, the *Municipal Water and Sewage Services Transfer Act*, transferred the ownership of 230 water and wastewater plants from the province's Clean Water Agency to municipalities and absolved the province of responsibility for constructing, expanding, or financing water or sewage works. One government document predicted that the bill would "prompt many municipalities to take a fresh look at their water and sewage management arrangements." Municipalities' increased fiscal restraint would also play a role: "This concern toward generating new savings will likely lead to more municipalities competitively tendering water and sewage management arrangements."[77] Others assumed that municipalities – which were tightening their belts as the province downloaded myriad financial responsibilities – would go so far as to sell their water and sewage assets. MPP Floyd Laughren predicted, "While municipalities may not, by nature, want to privatize, when the downloading hits them they will have no choice. . . . Once that squeeze really hits those municipalities they're going to be looking around every nook and cranny and under every stone to find a dollar in order to prevent outrageous tax increases. Guess what will jump out at them? It will be the sewer and water services in that municipality."[78] Toronto Councilor Peter Tabuns agreed: "In terms of water and wastewater systems, there will be pressure to sell those off to the private sector."[79] The London and District Labour Council's Gil Warren may have made the most sweeping predictions about the act's impacts, saying that "The sale of all water and sewer assets to large private corporations . . . is just around the corner."[80] The predictions proved groundless: Five years later, municipal asset sales remain virtually unheard of in Ontario.

Nor has the province privatized its own water and wastewater treatment business, the Ontario Clean Water Agency (OCWA). The Conservative government, which has for years paid lip service to the notion of the privatization of provincial services, renewed its commitment to the process in February 2000. "Everything is on the table, every idea will be considered, every concept will be explored," Finance Minister Ernie Eves vowed. "If the private sector can find a way of providing services currently provided by the government in a way that is more cost-efficient and improves the quality of that service, then we are ready to listen."[81] Its rhetoric aside, the government does not appear to be listening to proposals to privatize OCWA, a provincial Crown corporation that operates over 400 municipal water and sewage facilities serving 1.8 million and 3 million people, respectively, in over 200 municipalities. Although then Environment Minister Norm Sterling announced his intention to privatize the agency in 1996, and the Office of Privatization began a review of the agency in 1998, the proposal languished. It is commonly believed that "poison pills" in the agency's contract with the Peel Region rendered it virtually unsaleable.[82]

Ontario made another feint towards privatization in June 2000. A Cabinet document prepared by the Minister of Municipal Affairs revealed plans to instruct the province's 571 municipal governments to examine services in order to determine whether public or private provision would provide the best value. If governments could not prove that a public service provided better value, they would not be permitted to directly provide that service.[83] The press caught wind of these plans at the time the government was establishing a broad public inquiry into the Walkerton water tragedy. Sensitive, perhaps, to the inquiry's mandate to review the public and private sectors' roles in the provision of water services, Premier Harris denied any intention to force municipalities to examine privatization. "Nobody," he insisted, "is considering any development or privatization that I know of, of water or sewer."[84] Seven months later, however, the premier announced that SuperBuild Corporation would be hiring consultants to advise it on the restructuring of the province's water sector. The corporation would "see what kind of interest there is" within the private sector and explore private finance, design, construction, and operation options.[85]

Barriers to Privatization

Given the interest in privatization at several levels of government, the consensus among experts that privatization holds great promise, and the overwhelming evidence of privatization's benefits elsewhere, why do the provinces and their municipalities so often lose their courage when faced with a decision to privatize? Some are slowed by financial impediments, both real and imagined. A number of tax policies tilt the playing field in favour of the public sector.[86] The promise of government grants and subsidies also works against privatization, in so far as it reduces municipalities' needs for private capital.

Furthermore, municipalities commonly believe that public capital may be obtained more cheaply than private capital. Public borrowing costs may indeed be lower for municipalities with good credit ratings. The so-called advantage of public financing, however, fails to recognize the real costs of cheap money: the reduced credit ratings and increased future borrowing costs that may accompany higher levels of debt, the opportunity costs associated with using the capital for water and wastewater systems instead of other projects, and the assumption of financing risks by taxpayers rather than shareholders.[87] As economist Michael Klein has pointed out, interest rates are lower for governments not because they manage projects better and face lower risks but because they have recourse to taxpayers in case of project default. If the costs of the credit insurance that taxpayers provide for free were taken into account, he concluded, "it would no longer be clear that government credit is cheaper."[88] The Ontario government's Alternative Financing and Public-Private Partnerships Working Group went further in a 1997 report, placing lower financing costs at the top of its list of objectives for alternate financing. It explained that the ostensibly lower costs of public capital may actually hinder both a project and the province. The transfer of risks to the private sector, the preservation of the government's credit capacity, and lower construction, operation, and maintenance costs, it suggested, may overcome the cost barriers to private financing.[89]

In Ontario, the government created a financial impediment to full privatization by including in Bill 107 a provision requiring municipalities that wish to sell their facilities to the private sector to pay back all grants received since 1978. The grant repayment scheme fails to take into account the value of the facilities being sold. Conceivably, the grants could be greater than a facility's value – especially if the facility had since depreciated, or if the grants had been used to overbuild the facility, and if the facility served an insufficient number of customers to fully support its new, larger self. Repaying such grants could be economically unfeasible and could make the acquisition of the facility unattractive to a private firm.

The Ontario government has unwittingly discouraged privatization in another way, as well. It has propped up OCWA, giving it advantages over private sector firms and discouraging competition. The Canadian Council for Public-Private Partnerships, which represents both the public and private sectors, has expressed its concerns about "the apparent lack of a level playing field between the OCWA and private operators."[90] According to Eric Cunningham of United Water, OCWA is "the most serious impediment to the creation of a competitive environment in Ontario." He added that thoughtful companies will not compete against the agency.[91] Paul Boucher of Allied Water concurred. OCWA, he said, "is killing competition."[92]

OCWA's advantages include a generous provincial subsidy. When established, the agency acquired from the province, at a discounted rate, a municipal loan portfolio worth about $600 million. The interest rate spread between the loan portfolio and OCWA's loan from the province provides the agency with income that it has used to subsidize operations.[93] This subsidy prompted Stuart Smith, then president of Philip Utilities Management Corp., to observe, "you could be a chimpanzee and make a profit."[94] OCWA enjoys a host of other competitive advantages. It controls 95 percent of the outsourcing market in the province – a market share gained not through merit but through government fiat. It enjoys favourable tax status: It is exempt from charging GST or paying corporate taxes. Its balance sheet, the result of its subsidy, enables it to bid on large contracts. Furthermore, the agency is backstopped by the province for any financial liabilities. Perhaps as a result, in all cases but one, OCWA has been exempt from posting performance bonds. As the *Report on Business* explained, competitors believe that the province's financial guarantee allows OCWA to take risks and to act more aggressively than other firms.[95] A review by the Office of Privatization confirmed that the agency can appeal to municipalities because its contracts "are more flexible and assume more risk . . . than private sector contracts."[96]

Generally, it is politics rather than economics that gets in the way of privatization. Unions lobby hard for services to remain in the public realm. Many environmentalists likewise fight greater private sector involvement, often on the grounds that water is a uniquely precious substance that must not be tainted by the marketplace. Activists' aversion to the private sector's profiting from water treatment surfaced repeatedly at the Ontario Standing Committee hearings into the *Water and Sewage Services Improvement Act*, where speaker after speaker suggested that water's role in sustaining life should insulate it from market forces. "Water," insisted CUPE's Sid Ryan, "is an essential element in Canada's intricate web of life and not a commodity for corporate greed." Several private citizens agreed. "Safe, clean water is a necessity of life . . . ," said one, "not a commodity that should be delivered to our taps at a price that will fatten the pockets of private investors." Another echoed these concerns: "Because of the fundamental human need for clean water, . . . water resources should not be subject to market forces."[97] No one thought to ask why we trust the production and distribution of food – so essential to sustaining life – to private farmers and grocers.

Decision makers may also be spooked by the spectre of water exports. "When the same companies setting up shop in Ontario are also acquiring water contracts in the U.S.," asks the Canadian Environmental Law Association, "can water pipelines to the U.S. be far behind?"[98] Of course they can be very far behind. Private firms acquire the right to serve customers in a given area; they do not acquire an unlimited right to extract water from a given source for whatever purpose suits them. But the confusion is understandable. Amanda Martin, then president of Azurix, the Enron company that operated both water utilities and an Internet water exchange, told an interviewer of her firm's interest in gaining access to Canadian water. "There'll be increasing pressure on Canada to export water," she said. "We believe there is a market in the U.S. that can be served by Canadian water supplies."[99] Little wonder that CUPE's Judy Darcy warns, "Multinational corporations are trying to privatize water services in hundreds of Canadian municipalities and turn our water resources into an export commodity. They can't buy the air we breathe, so now they want to buy and control the water we drink."[100]

In 2001, opponents of privatization fabricated a new bogeyman. They warned that international trade agreements could limit municipalities' powers to control privately operated waterworks. CUPE commissioned a legal opinion from lawyer Steven Shrybman suggesting that the North American Free Trade Agreement (NAFTA) and the World Trade Organization's General Agreement on Trade in Services (GATS) could hinder attempts to introduce more stringent regulation or to reverse a decision to privatize.[101] The federal government countered that fears that trade agreements could threaten local autonomy over water quality were unfounded.[102] One trade lawyer dismissed the Shrybman opinion as "one of the most extreme and alarmist legal interpretations of trade agreements I have ever read."[103] Another pointed to glaring errors, calling it an "absurd" analysis that "ignores reality" and concluding, "Contractual provisions can easily deal with all of the issues that concern municipalities entering such partnerships."[104] A law firm retained by the Canadian Council for Public-Private Partnerships roundly challenged Mr. Shrybman's claims. No provisions of NAFTA, it insisted, inhibit governments from enacting water quality standards, imposing operating conditions on a utility, or returning a service to the public sector after it has been contracted out. A properly drafted contract can eliminate the application of NAFTA with respect to expropriation. As for GATS, it does not apply to water-supply contracts. It concluded, "NAFTA and other trade agreements are not the threat that the Shrybman Opinion would have us believe. Those agreements provide a moderate level of protection to investors and their investments while maintaining municipalities' freedom to enact measures designed to protect the interests of their citizens."[105]

Given the barrage of anti-privatization "information" – however inaccurate and irrational – Canadians have been exposed to, it is hardly surprising that many would develop concerns about the process. The Canadian Union of Public Employees claims that, when polled in 1998, Canadians, by a margin of five to one, preferred publicly owned to privately owned water utilities.[106] The Canadian Environmental Law Association and Great Lakes United report the results of another poll: When asked who should control water systems, 76 percent of the respondents said municipal officials.[107] Although these polls did not test public support for publicly controlled, privately operated systems, many agree that the public harbours concerns about privatization and that governments' fears of adverse public opinion may be barriers to successful partnerships.[108]

Another factor preventing governments from embracing privatization is simply the fear of the unknown. Municipalities have little experience. They want more information on what works and what does not. They want case studies, templates for successful contracts, and models for creating and managing partnerships and assessing their success or failure. As their counterparts both within and beyond Canada's borders accumulate and share experience, they will doubtless gain confidence. Ultimately, as one provincial official noted, they will just have to take the plunge: "The public sector needs to jump in the pool and try it. They are looking too long. They need to be bolder. Part of the learning process is making mistakes."[109]

CHAPTER 6

TESTING THE WATERS
Early Experiments with Privatization

While privatization remains rare, Canadians are taking steps in that direction. In 1998, the Canadian Council for Public-Private Partnerships (CCPPP), the Canadian Chamber of Commerce, and the Federation of Canadian Municipalities sponsored a national survey of 211 players, including 34 provincial government representatives and 97 politicians and administrators from 57 municipalities with over 40,000 residents. Of the municipal and provincial respondents, 5 to 15 percent expected to see new or expanded partnerships for sewer construction or water supply within two years.[1]

The same year, CCPPP inventoried planned or completed public-private partnerships across the country. Although the inventory listed over four dozen water and wastewater projects, many of them did not live up to their promise as privatizations, having been cancelled or having involved contracts to public rather than private firms. Several of the larger projects remain little more than a gleam in a council's eye: Montreal's thoughts of privatizing its water and wastewater systems have remained half-formed for years and Winnipeg's consideration of privately built and operated water supply and treatment facilities has progressed very tentatively. Furthermore, a number of the projects described comprise just one component of a larger system. For example, Kelowna, B.C., has a five-year contract with a private firm to finance, supply, install, maintain, and read water meters. Winnipeg contracts out two services related to sewage treatment: the agricultural spreading of biosolids and the generation of oxygen. The Region of Niagara has a five-year contract for the management and disposal of sludge from its eight wastewater treatment plants; the private firm not only hauls the sludge but also handles the environmental permits, soil testing, and marketing to farmers.[2] Despite these qualifications, the council's inventory clearly pointed to growing private-sector involvement in the water and wastewater industry, predominantly in Ontario and Alberta.

The contracts described by CCPPP have generally brought specialized skills and savings to communities. They have less often brought access to private capital, in part because, as noted in the previous chapter, many municipalities believe that public capital may be obtained more cheaply than private capital. Several contracts have involved a mix of public and private financing. As noted below, public money financed 41 percent of Alberta's CU Water project. The Manitoba Water Services Board provided 50 percent of the capital for the water treatment and supply system for Cartier, Headingly, Portage la Prairie, and St. Francois Xavier. Grants from the Canada/Ontario Infrastructure Works Program financed two-thirds of a pumping station and sewage collection station for the Eastern Ajax area. In at least one case – the expansion of a wastewater lagoon for the hamlet of Langdon, Alberta – a private developer assumed the financing risk for a project on the condition that it would be repaid if subsidies could be secured.[3]

It is likely only a matter of time before private financing becomes commonplace. Moncton, New Brunswick's, new drinking water plant – discussed below – was completely privately financed. USF Canada, the firm that won that job, is confident that, with its parent Vivendi, it can meet the capital needs of any Canadian municipality.[4] United Water Canada, backed by Suez, likewise assures prospective clients of its financial capabilities: "With combined revenues of over $40 billion worldwide and over $3 billion in assets already under management in North America, United Water has the financial resources necessary to finance capital improvements."[5]

Regardless, communities have thus far appeared to be more interested in operations and maintenance contracts than in private financing arrangements. Forty-two Ontario communities have entered into water or wastewater agreements with private operators.[6] Typical arrangements involve three-to-five year contracts for operating the sewage treatment plants of small communities such as Forest, Listowel, Petrolia, and Plimpton. Tiny communities can also work with private firms, as is evidenced by the contract to finance, design, build, and operate a sewage system for Campden, a hamlet with just 80 homes.[7] Rarer are contracts that cover many facilities spread out over a large geographic area. One notable exception is the contract signed in 1998 by the Regional Municipality of Haldimand-Norfolk and the Professional Services Group, which covered seven wastewater treatment plants, six lagoon systems, and 43 pumping stations serving 46,600 people living in a 2,800-square-kilometre area.[8] Also rare are contracts for systems serving larger communities, such as the 400,000-strong City of Hamilton – a contract analysed in detail in the following chapter.

In September 2001, Azurix North America – since sold to American Water

Works – won a 10-year operations and maintenance contract for the water supply systems serving London, Ontario, and 20 communities in the surrounding region.[9] Six teams initially competed for the contract, and four were short-listed. The winning firm promised annual savings of approximately $1 million. Although a competitor had offered greater savings, it had been unwilling to provide the unlimited liability demanded by London.[10] Although disappointed by its bid's lack of success, one competitor did find a silver lining: It was delighted that the winner had succeeded in snatching the contract away from the incumbent Ontario Clean Water Agency.

The Ontario town of Goderich – population 7,500 – entered into a water and wastewater operations and maintenance contract with USF Canada in December 2000. The town had received eight submissions in response to a request for qualifications, and had weighed four proposals. The five-year contract, renewable for another five, will bring savings of over $71,000 a year.[11] While the decision ultimately rests with the town council, town administrator Larry McCabe assumes that savings will be used to benefit the water and wastewater systems.[12] The savings largely reflect a reduction in staff from eight to six, made possible in part by integrating the operations of the water and wastewater systems, which had previously been split between the local Public Utility Commission and the municipality. In keeping with council's concern for its employees' welfare, no staff were laid off: One was re-deployed to the electric utility and another took early retirement. All of the transferred staff were given equal or better wages and benefits.[13]

Savings, while welcome, did not drive Goderich's decision to privatize. "The primary objective," mayor Delbert Shewfelt explained, was "to improve safety and reduce the risk of harm to our residents, and we feel that a public-private partnership best accomplishes these goals."[14] The contract enhanced the accountability of the water utility; the mayor noted that for the first time, the town had a document specifying what the water utility should do. And if the town were for any reason not fully satisfied it could, with 180-days notice, terminate the contract without penalty.[15] The arrangement also promised improvements in service – in particular, a reduction in the number of bypasses at the sewage treatment plant – and in computer-enhanced equipment maintenance.[16] The town looked forward to taking advantage of USF's – and parent Vivendi's – expertise. The deal gave the town access to state-of-the-art management systems along with technology at below-market costs.[17] It also gave it an operator who cannot afford to make mistakes. In the words of the mayor,

"The chances of getting in trouble decrease because they have a huge reputation to live up to."[18] The mayor welcomed the transfer not only of responsibility for treatment but also of "some of the liabilities," noting the inherent hazards of providing water in an agricultural area. The Walkerton tragedy, it seems, had reminded the town of its own potential vulnerability.

A number of Alberta communities are also experimenting with contracting out the operations and maintenance of their water or sewage systems. In its 1998 inventory, CCPPP described five such arrangements in the province. Among the contracting agencies was the Alberta Capital Region Wastewater Commission, which has contracted out the operations of some of its facilities for more than 15 years. In 1998, it awarded an eight-year contract to OMI Canada for the operation and maintenance of one sewage treatment plant and five pumping stations serving approximately 150,000 people in 12 municipalities surrounding Edmonton. The contract brought savings of approximately 10 percent in the first year and 18 percent in subsequent years. The savings will average $400,000 per year.[19] Further south, eight firms contended for a contract to improve Banff's sewage system. The resort town needed to improve effluent quality and to expand its treatment capacity in order to accommodate the tourists that crowd it each summer. In 2001, Banff signed a contract with Earth Tech Canada to upgrade its treatment plant and to operate and maintain it for 10 years.[20]

Several communities outside Edmonton took a different approach to their water supply system. In the early 1990s, Tofield, Ryley, and their neighbours wanted to pipe treated water from Edmonton. Their county approached CU Water Ltd., a division of a natural gas company that owned a right-of-way along the highway and could thus build a pipeline without spending time or money acquiring land or easements. Two years of negotiations and a plebiscite gaining public support for the project followed. CU agreed to design, build, own, operate, and maintain a pipeline in exchange for an exclusive 25-year franchise. It also agreed to finance $7.1 million of the construction costs, with the province providing the balance of $4.9 million. The agreement entitles the Alberta Public Utilities Commission to buy back the system at net book value at the 15th, 20th, and 25th years of the agreement, with a five-year notice period. Ryley administrator Bob Luross called the deal a godsend, adding, "It's been a big load off our minds. Our treatment plant was 40 years old and it was a big chore to keep up. Now they run everything and we are out of the water business."[21]

Privatization has long had a toe-hold in British Columbia, where 187 privately owned water utilities serve approximately 30,000 households in the province.[22] More than half of these utilities are very small, serving fewer than 50 customers in trailer parks, resort areas, subdivisions, or isolated communities. The largest – White Rock Utilities – has been operating since 1913 and supplies 18,500 people.[23] In recent years, operations and maintenance contracts appeared to be gaining momentum in the province. Victoria contracted with a private firm to treat septage from approximately 100,000 people.[24] And the Greater Vancouver Regional District (GVRD) planned to engage the private sector in the design, construction, and operation of the Seymour water filtration plant. The plant – the district's first filtration plant – was to remove giardia and cryptosporidium cysts, reduce bacteria and organic matter, and eliminate turbidity in order to make disinfection more effective, ensure compliance with provincial and federal standards, and improve the appearance of the water.[25]

GVRD issued a request for qualifications in the fall of 2000 and, in February 2001, announced a shortlist of four consortia from which it would invite full proposals. It expected to issue a request for proposals later in the year, to award a contract in 2002, and to see the plant completed in 2005. It envisioned a contract with a 20-year operating term. According to lead engineer Mark Ferguson, involving the private sector was "all about efficiency." The district had simply found public-private partnerships to be more competitive than purely public alternatives. Councillor Marvin Hunt, chairman of the GVRD water committee, anticipated that privatization would bring savings of between 15 and 20 percent. Private operations have elsewhere "been extremely successful," he enthused. "There's significant monies there. I am very much in favour of the concept. It's the best way to deliver the safest product for the least amount of money for the taxpayer."[26]

In the end, opposition spearheaded by the Canadian Union of Public Employees (CUPE) and the Council of Canadians defeated the proposed plan. Opponents, trumpeting their fears that international trade agreements would enable a private operator to wrest control from public regulators, mobilized more than 1,000 protesters to show up at public meetings. The atmosphere of fear and distrust proved to be too much for councillors to bear. Opponents had sent "a very clear message," Councillor Hunt acknowledged. "We said we would listen. And we did."[27]

CUPE likewise spooked the councillors in Kamloops, BC, who in July 2001 rejected a privatization option for their city's drinking water.

Kamloops, whose turbid water frequently fails to meet Canadian drinking water guidelines and whose residents are periodically advised to boil their water or to purchase bottled water, had explored a private solution to its problems. The city retained PricewaterhouseCoopers to study options for a new water treatment plant. The consultant acknowledged that selecting a private firm would take time and could lack strong political support. Nonetheless, it recommended that the private sector design and build the plant and operate it for 20 years. It also recommended private financing as the best way to reduce the city's risks and risk-adjusted costs: "If the full benefits of risk transfer are recognized and greater affordability is desired, the preferred delivery model for the plant would be Design, Build, Finance, Operate."[28] Labour fought back with presentations to city council stressing, among other issues, Vancouver's change of heart on privatization. The arguments had their intended effect. Councillors, expressing concerns about trade agreements, project delays, irreversible commitments, and community opposition, voted against entering into a private-public partnership for the operation of the treatment plant.[29]

CUPE worked equally hard, but this time unsuccessfully, to defeat privatization on the other side of the country, in the regional municipality of Halifax, Nova Scotia. In that province, where the need for $572 million in water and wastewater projects has been identified, the promise of private financing attracted the cash-strapped government as early as 1994.[30] "I see a big, big role for the private sector," said Economic Development Minister Ross Bragg. "Government can't go out and finance all these things themselves. Why not let the private sector do some of these sewer and water projects?"[31]

Why not indeed, wondered Halifax, which proposed the private design, construction, and operation of several new sewage treatment plants in order to stop the centuries-old practice of dumping raw sewage – currently, more than 43 billion litres a year of it – into its harbour. Scientists issued warnings about the harbour pollution as early as 1924, and an array of politicians, committees, commissions, task forces, and panels considered solutions for almost as long, but the costs of cleaning up proved too daunting.[32] In 1996, the municipality renewed its determination to find a solution. The prospect of over $200 million in private investment persuaded it to explore private options. It tested the waters in 1998 with a request for expressions of interest that drew 22 responses, convincing it that the private sector had sufficient capacity to undertake all aspects of the project, including financing.[33] Although Halifax later decided that public financing was the less expensive option, it was prepared to finance only two-thirds of the capital costs; it solicited federal and provincial funding for the balance. Halifax received final proposals from two private consortia and a reference bid from municipal staff. In October 2001, the municipality's selection committee revealed that the preferred bid had been submitted by a consortium led by United Water Services Canada. Six weeks later, despite CUPE's having lobbied hard to derail the project and the mayor's having stated his preference for public operations, council agreed to start negotiating a contract with the consortium to design, build, and operate three treatment plants.[34] It quelled any lingering unease by securing the option to take over plant operations in years six, 12, 15, or 20 of the 30-year agreement, albeit not without financial penalties. Council approved draft agreements in May 2002. If environmental approvals and financing materialize as planned, the new collection systems and treatment plants will be up and running by 2007.

Private involvement in the project will bring considerable savings. At $262 million, the consortium's capital costs are almost $53 million less than Halifax's initial estimate.[35] Differences in the makeup of those figures, however, make precise comparisons between the two impossible.[36] Perhaps more meaningful is a comparison of the consortium's estimates of capital expenditures and operating costs over the life of the project to those contained in the municipality's reference bid. The respective net present values of $465 million and $479 million suggest that the consortium will save Halifax $14 million.[37] That figure underestimates the financial benefits by ignoring the different assignments of risk. The consortium's estimates are guaranteed by water giant Suez; those in the reference bid are backed by Halifax taxpayers.

In celebrating council's decision to negotiate the contract, the editors of the *Halifax Herald* wrote, "They certainly can't be accused of breaking any land-speed records, but Halifax regional councillors still deserve credit for finally making a decision to move ahead on cleaning up the raging pollution in Halifax Harbour. . . . Visitors to the waterfront and boaters have suffered from this disgusting effluvium for long enough. The environmental degradation must stop."[38] Thanks to council's decision to involve the private sector, "this disgusting effluvium" will soon be a memory, and the region's dream of a clean harbour will be closer to becoming a reality.

The East coast is also home to Canada's largest completed private drinking water project. USF Canada operates and maintains the water filtration plant that it financed, designed, and built in Moncton, New Brunswick. Prior to the plant's construction, the city had struggled with discoloured, bad tasting, sub-standard water for many years. High bacteria counts in the municipal water system had led to several boil-water orders, including one in 1997 that stretched for 36 days.[39] The following year, the city's water failed to meet standards for pH, turbidity, and total tri-halomethanes.[40] At one point in 1999, contamination with coliform bacteria prompted the city to haul clean water in stainless steel tanker trucks from another town's treatment plant.[41] One reporter marvelled over city council's decision to solve the problem by constructing its own filtration plant: "Imagine drinking water right out of the tap and enjoying it. For most Greater Monctonians, that concept is as foreign as some of the substances they have found in their water over the past couple of years."[42]

Unable to obtain provincial or federal funding for a water treatment system, Moncton turned to the private sector for help. Ron LeBlanc of the city's engineering department explained that working with the private sector was the only way the city could afford to construct a state-of-the-art system.[43] In 1998, after a competitive bidding process that initially saw expressions of interest from nine consortia, the city signed an agreement with Greater Moncton Water, a company owned by USF Canada and the Hardman Group (the latter of which was later bought out by the former). The company offered considerable expertise: Parent United States Filter manages more than 500 facilities in North America and, as noted above, is a subsidiary of the French water giant Vivendi, which has operations in more than 100 countries.[44] Moncton is delighted to gain access to the company's patented technologies, computerized systems, and resulting efficiencies. In the words of City Manager Al Strang, "We came up with a far superior deal than if we could have built it ourselves."[45]

Privatization brought immediate financial benefits to Moncton. The arrangement relieved the city of having to make any up-front capital investment. Equally important were the substantial cost savings. Greater Moncton Water built the plant for $23 million – between $8 million and $10 million less than a publicly designed and built plant would have cost. Those savings resulted in part from a 40 percent reduction in the size of the building, which was made possible by the choice of a particular kind of filtration. Operating costs will also be lower than they would have been at a publicly run plant. All told, the city expects to save between $14 million and $17 million in capital and operating costs over the course of the 20-year lease; estimates of savings have ranged from $12 million to $20 million. The city will pass along these savings to consumers. The average household will pay $91 a year for the plant instead of the $119 anticipated under the public alternative. Mr. Strang expressed his pleasure with the deal, calling both the city and the company "winners."[46]

Moncton's privatization also brought dramatic health benefits. In an editorial congratulating the city for its decision to proceed with a public-private partnership, the local newspaper enthused that residents "should be able to celebrate clean, clear, and contaminant-free drinking water for the first time in recent memory."[47] The contract requires the operator to meet or exceed Canadian drinking water guidelines. Its requirements for aluminum and colour are considerably stricter than the guidelines.[48] The contract also specifies turbidity levels of less than 0.1 nephelometric turbidity unit (NTU) at all times. This significant improvement over the average 1.87 NTUs recorded in 1998 will improve the taste and smell of the water and reduce chlorine requirements and subsequent trihalomethane formation.[49] Mr. LeBlanc boasted, "We believe the water that we have specified will be the best in Canada," adding, "If they don't meet the specs, then they ain't getting paid."[50] The new plant quickly lived up to its promise. After four months in operation, trihalomethanes had been reduced by almost 75 percent and chlorine consumption had been reduced by more than 70 percent.[51]

Moncton's satisfaction with its private water plant prompted it to look at other privatization candidates, among them, the city's water and sewer pipes.[52] Rehabilitating the distribution and collection system, some of which is 122 years old, could cost the city as much as $70 million and could take up to 20 years. In June 2001, Mayor Brian Murphy speculated about the benefits of involving the private sector in the process: "If it can be demonstrated that a private sector firm can do it better, faster, and cheaper, all while keeping control in the hands of council and protecting the jobs, salaries, and benefits of our employees . . . Stay tuned!"[53]

In its enthusiasm for its water provider, the city may have become incautious regarding process. As it worked with USF on an agreement to rehabilitate, operate, and maintain the distribution and collection network, it presented the work as an extension of the existing contract. It noted that the original tender for the filtration plant had asked bidders if they would be interested in working on the distribution system at a later date. It also argued that the treatment and distribution facilities should be run by the same company. But others were not convinced. A competitor complained loudly of being excluded from bidding on the project, while interest groups emphasized the need for greater transparency and public involvement.

In March 2002, Moncton city council decided to seal the draft agreement with USF and commission an independent study of the needs of the system and the costs of upgrading it. The study, expected to take a year to complete, should give the city a baseline against which to measure the USF proposal and help it determine whether to proceed with the agreement, perform the work itself, or initiate a competitive bidding process.[54]

CHAPTER 7

WHAT WENT WRONG?
Water and Wastewater
Privatization in Hamilton, Ontario

The contracting out of water and wastewater operations in Hamilton, Ontario, has earned decidedly mixed reviews.[1] Five years into the contract, Art Leitch, then General Manager of Transportation, Operations, and Environment for the Regional Municipality of Hamilton-Wentworth, praised the arrangement. "Not only do operating costs continue to decrease," he maintained, "but Hamilton is realizing extensive private capital investment in the community and economic development benefits as a result of this successful public private partnership."[2] The previous year, Canadian Union of Public Employees (CUPE) president Judy Darcy offered a wildly different assessment: "This is the worst example, bar none, of the horror stories we've heard about privatization."[3] It is hardly surprising that Canada's largest water and wastewater privatization to date would provoke intense criticism from labour unions. But the unions bolster their arguments with studies by several academics. And at least some of their concerns are echoed by policy analysts normally favouring privatization. A critical examination of the arguments on all sides reveals that Hamilton's non-competitive approach to privatization has brought moderate savings and investment but has not yet solved many of the severe problems plaguing the city's systems.

Sewage pollution has disgraced Hamilton for many decades. Since the 1940s, when authorities first closed local beaches, through the 1980s, when the United States and Canada declared Hamilton Harbour an "Area of Concern" under the Great Lakes Water Quality Agreement, to the present day, an inadequate sewage collection and treatment system has soiled the city's waters and reputation. A series of problems have beset the main sewage treatment plant at Woodward Avenue. For decades, inadequate sampling allowed the plant to exceed pollution limits without detection up to 15 percent of the time.[4] Even so, it and Hamilton's two other sewage treatment plants made eight appearances on the province's lists of plants not complying with environmental standards between 1987 and 1994. Tight budgets forced the postponement of costly upgrades. The plant's managers mis-spent many of the funds dedicated to the system. In the 1980s, in what came to be know as "Sludgegate," the region wasted $13 million on a thermal kinetic sludge-drying system that never worked – a mistake that then-Councillor Dominic Agostino called "probably the largest screw-up . . . in the history of the region."[5] A bloated staff at the plant ate up millions more. Then-Councillor Dave Wilson explained that politicians could not muster the will to downsize. The plant, he recalled, "was the absolute worst political football in the region. Whenever managers down there tried to make something happen, they were typically prevented from doing so by the politicians."[6] His candid assessment: "We never ever were able to manage the plant as successfully as we should have been."[7]

Indeed, the early 1990s saw one scandal after another at the Woodward Avenue plant. One of the plant's most dogged critics, columnist and former councillor Michael Davison, described the plant as "the region's most screwed-up facility"[8] and one "that has caused the politicians considerable embarrassment that is likely to get worse."[9] In column after column, Mr. Davison documented poor communication between the plant and its political bosses, sagging morale, lax security, personal use of regional equipment and staff time, and two sex scandals, one involving a stripper performing in the plant's lunchroom and the other involving almost $2,000 worth of telephone sex calls. Councillor Agostino likely spoke for many of his colleagues when he complained, "I've some very major problems with and concerns with the way that place is being managed."[10]

Regional councillors must have felt relieved when, in April 1994, a firm approached them with an unsolicited proposal to operate their water and wastewater system. Yet the attraction of turning over to another party a nagging management headache or the prospect of solving long-standing operating problems were not their primary reasons for pursuing the proposal. Above all, the councillors saw in the offer an opportunity for economic development. The proposal came from Philip Utilities Management Corporation (PUMC), newly formed by the Hamilton-based waste management giant Philip Environmental Inc, later known as Philip Services. PUMC needed experience under its belt in order to break into the North American water and wastewater business. While it could not offer Hamilton expertise in the field, it could promise to bring economic development to the region.

Visions of becoming a thriving international waste and water treatment centre swept council away. Philip Environmental promised that it or related companies would create 100 full time jobs in the region over five years; in the event it failed to do so, it promised to pay the region $10,000 for each job not created. It also agreed to spend at least $15 million on new capital projects in the region and to develop an environmental enterprise centre. PUMC's own commitments included promises to locate its head office in the region, to build between 15,000 and 25,000 square feet of new office space to accommodate it, and to establish an international training centre with a local college.[11] Councillor Wilson remembers, "It was going to make us well-known across the world as the headquarters, and it would be good for the corporation here and obviously good for Hamilton in general."[12] Bureaucrats also believed that PUMC's presence would attract similar businesses to the region, since industries, at least according to Mac Carson, then Chief Administrative Officer for the Hamilton-Wentworth region, tend to "nest," or to locate in a common area.[13]

The local press was equally enthusiastic. "Cash may soon flow from sewers," gushed the headline of an article forecasting the creation of hundreds of jobs.[14] A *Hamilton Spectator* editorial called the proposed deal "an impressive economic opportunity."[15] Columnist Jack MacDonald went further, calling the proposal "a once-in-a-lifetime opportunity" and "a brass-ring chance to attract world market."[16] A magazine, *Ecolutions*, sprang up to chronicle the rise of environmental industries in the region. As one newspaper reporter recalled, "It all sounded so promising."[17]

Key to Hamilton's economic development dream was the aiding of a local company. Few, if any, councillors saw PUMC's lack of expertise in operating water and wastewater systems as a hindrance. Indeed, one purpose of the contract was to allow the young company to gain the experience it so clearly lacked. Council explained, "In order to compete in the lucrative international environmental services market, PUMC needs to establish its credentials by operating wastewater [sic] and sewage treatment plants."[18] Leo Gohier, then Director of Water and Wastewater for the region, confirmed that council wanted to give PUMC a leg up: "The basis of the negotiations was essentially to develop a contract that would allow a local firm to develop expertise and experience in the contract operations of a major municipality's water and waste water facilities, while providing the region with a vehicle for economic development."[19] Councillor Wilson elaborated on the region's hopes for PUMC: "The theory was that by letting them manage the plant, . . . they would be able to show others what a great job they could do operating the plant and the system and eventually they would be able to sell their skills right around the world."[20]

PUMC's political credentials may have contributed to council's enthusiasm for the firm. The company's founding president, Stuart Smith, had led the Ontario Liberal party as the Member of Provincial Parliament for

Hamilton West. In the late 1970s, Dr. Smith had employed Sheila Copps, who went on to become Hamilton's federal parliamentarian, serving at the time the PUMC contract was let as both Deputy Prime Minister and Minister of the Environment. The *Hamilton Spectator* credited the charismatic and connected Dr. Smith as "the single greatest influence selling the deal to politicians."[21] Some have wondered if politics exerted undue influence in the awarding of the contract.[22] Others have defended the choice to work with a familiar party. As Mr. Carson explained, "the region felt comfortable with the people on the opposite side of the table . . . [T]his was another important factor in determining whether sole sourcing would be an effective model."[23]

Given council's determination to work with PUMC, competition for the contract was out of the question. At its April 19, 1994, meeting, council voted unanimously to enter into exclusive negotiations with PUMC. "The reason it wasn't tendered is because it was an economic initiative as opposed to a straight operational issue," recalled Councillor Wilson. "If it was simply our desire to have the plant operated by a private corporation, it would have been tendered."[24] Not surprisingly, at least one would-be competitor complained of being shut out. Wheelabrator claimed that it could offer substantially greater savings than could PUMC.[25] But the region was not after greater savings. In Mr. Carson's words, "the region was not looking to get every last cent, but wanted a local company committed to the objectives and to creativity."[26]

It was no secret that, without competition, the region might not get the best possible deal. In order to protect its interests, it hired KPMG Management Consulting to, among other things, evaluate the fairness of the proposal. In the absence of competitive bids that would provide the only truly meaningful comparative reference points, KPMG could not determine that the arrangement was the best possible. However, after reviewing 14 similar projects south of the border, it concluded that the agreement would fall "within the range of outcomes that could reasonably be considered to be 'fair' to the region." This lukewarm endorsement appeared to trouble no one. As KPMG partner Will Lipson explained, "sole sourcing may not be the way to get the absolute best deal, but it is a way to get a fair deal that's good enough, that's a win-win situation and addresses the criteria that matter the most, which in this case revolve around economic development." Liam Rafferty, then the region's assistant corporate counsel, likewise defended the possibly inferior deal: "It seems not improper for a council like ours, when faced with a proposition that is a good deal, although possibly not the absolute best deal, to look seriously at it. . . . [E]ven if a better deal was possible (which is doubtful) the council would probably have selected the local guys with the good deal over the out-of-town guys with the better deal."[27]

PUMC's Dr. Smith also championed council's decision not to tender the contract. Dr. Smith has in another context praised open competitions for contracts, calling them "very valuable as a source of savings to Ontario's municipalities." "Competition works," he told a provincial legislative committee in 1997, "and I think most of us know that, but it's been proven over and over again. The truth is the competing firms are obliged to be as innovative as they possibly can be."[28] Regardless, Dr. Smith did not want to see competition in Hamilton. In defending the sole sourcing process, he tacitly admitted that his company was less attractive than others in the field. "It was absolutely proper," he maintained, "because there was no way in the world a Canadian operator could have won."[29]

By the end of the year, the regional municipality, PUMC, and Philip Environmental had signed two contracts, effective January 1, 1995: a 10-year contract to manage and operate one water treatment plant and three wastewater treatment plants and a renewable two-year contract to maintain and operate outstations and the high lift pumping station for the water treatment plant. In addition to committing PUMC and its parent to their economic development promises, the contracts spelled out economic arrangements, established performance criteria, and addressed labour issues.

Under the plant operations agreement, the region agreed to pay PUMC an annual fee equal to the $18.6 million it had previously budgeted to run the plant, less $500,000 in guaranteed savings for the region, $103,000 to cover the environmental services department's overhead, and $100,000 for contract co-ordination.[30] Subject to various adjustments, the contract assigned 10 percent of further cost savings to system employees and the next $1 million in savings to PUMC. It divided additional savings between PUMC and the region, with 60 percent going to the former and 40 percent to the latter. Should savings reach 20 percent of the annual budget, PUMC's share would rise to 80 percent.[31]

The contract required PUMC to maintain the levels of service previously achieved by the region. PUMC provided a $5-million performance bond to guarantee its performance (except in the area of economic development) along with $20 million in both general liability insurance and pollution legal liability insurance. To further protect the region, Philip

Environmental guaranteed PUMC's performance. Furthermore, the region could terminate the contract if PUMC's operations threatened health or public welfare and the firm failed to remedy the situation within two business days of written notice.[32]

The contract also addressed labour issues, incorporating agreements that PUMC had signed with the two unions affected by the deal. PUMC promised that it would abide by collective bargaining agreements, that it would not lay off any of the system's employees until March 31, 1996, that it would increase rates of pay in January 1995 and again in January 1996, and that it would share a portion of savings with employees. It also agreed to establish a pension plan with equal or better terms than the public pension plan; in the interim, PUMC's employees would be considered to be on extended leave of absence from their municipal positions, allowing them to remain in the public pension plan.[33] In wooing the workers, PUMC offered them 15 percent of cost savings achieved; it ultimately agreed to a 5 percent wage increase and a 10 percent cost savings arrangement.[34] PUMC also held out hopes that its future international expansion would lead to new job opportunities. Satisfied by these promises, the International Union of Operating Engineers (IUOE), which represented the bulk of the system's employees, endorsed the privatization. IUOE's business manager Peter Yemen, admitting that "privatization is coming whether we like it or not . . . [W]e're not going to stop privatization," explained, "we've got the opportunity to get in now and work with this company, and we should make the best of that."[35]

Success or Failure?

Six years into the contract, its successes and failures continue to be vigorously debated. Considerable disagreement exists concerning Philip's and PUMC's realization of the contract's economic development goals. The former's financial, environmental, and legal difficulties left it a shadow of its former self. PUMC, however, did grow. By March 1999, when it was sold to Enron's Azurix Corporation, it operated 14 municipal treatment facilities – mainly in small communities – in North America and claimed approximately $130 million in material assets.[36] More impressive, it had won a $101-million, 25-year contract to design, build, and operate Seattle's Tolt water treatment facility. Dr. Smith called winning the Seattle contract an "important achievement," boasting, "It showed that Canadians could compete with the best in the world, which was the whole point of the [Hamilton] exercise."[37]

However, the extent to which PUMC's success rubbed off on Hamilton remains controversial. In a 1999 editorial, the *Hamilton Spectator* charged that promised economic benefits had never materialized and concluded, "For their noble intentions in this matter, councillors got played for patsies."[38] Others disagreed. Fred Eisenberger, then chairman of the region's environmental services committee, maintained that the deal benefited the region, although not as directly as anticipated.[39] Terry Cooke, then regional chairman, admitted that there was "room for improvement" but said that Philip Services and PUMC, with their investments in jobs and their success in "attracting interest to the community," had "fulfilled the spirit and generally the substance of the commitments that were made."[40]

Neither critics nor defenders dispute that PUMC lived up to only some of its economic development promises. Instead of constructing new office space, the firm, with the region's blessing, refurbished several floors of an existing building in downtown Hamilton.[41] By the end of 1998, PUMC had made only $6.5 million of the promised $15 million in capital investments in the region.[42] Nonetheless, when the firm was sold to Azurix, the region agreed that it had fulfilled its obligations in this regard and that no further action would be required.[43] At the time of the sale, the firm had neither developed an environmental enterprise centre nor established an international training centre; Azurix assumed its commitment to doing so.[44] Job creation promises remain difficult to assess. While parent company Philip Services created some jobs, its own restructuring and subsequent move to the United States, along with PUMC's staff reductions at the Woodward Avenue plant, destroyed others. The region, however, was satisfied. In its review of PUMC's performance throughout 1998, it noted that more than 100 jobs had been created in various operations.[45] Upon approving PUMC's sale to Azurix, the region confirmed that the firm had fulfilled its job-creation obligations.[46] Azurix also strengthened its local employment numbers, moving an engineering division from nearby Burlington to Hamilton.

The change to American ownership prompted some to question the success of Hamilton's plan to nurture a local company. Dr. Smith wondered if Azurix would use Canadian skills and technology. "I am unhappy about what happened," he mused, "because I fear that the main purpose of the Hamilton contract may have been defeated. The idea was to have a major export industry in water and sewage services, based in Canada. In fairness, perhaps Azurix still intends that and they may have a mandate from Enron to do so. Have they succeeded?"[47] The sale of Azurix, announced in August 2001, to the long-established New Jersey-based American Water

Works suggests a negative answer to that question.

While others have expressed reservations, the operator has steadfastly maintained that the contract has brought considerable economic benefits to Hamilton. In June 2001, Azurix reported that it had created 117 new jobs in Hamilton, bringing an additional $5.7 million in annual salaries and many direct and indirect benefits. It valued its creation of jobs at $95.8 million over the contract period. In addition, it predicted that the city would enjoy new revenues from a carbon partnership along with operating savings and avoided capital investments. In all, it maintained that the total economic and contract benefit would be $130.8 million in nominal dollars.[48] The city seems to have been persuaded: Four months later, a report prepared by Hamilton water quality manager Jeff McIntyre estimated that Azurix pumps $40 million a year into the city economy.[49]

Although the promise of economic development was the primary factor in council's decision to privatize, it was not the only factor. The partnership, council also anticipated, would "provide environmental benefits."[50] Those benefits have proven somewhat elusive. Hamilton Harbour remains an international Area of Concern. The September 1998 status report on the Remedial Action Plan for the area noted unacceptable levels of ammonia, phosphorus, and suspended solids in the harbour and scolded the Woodward Avenue sewage treatment plant for not meeting its targets.[51] Sewage pollution continues to curtail swimming. Between 1995 and 2000, swimming was permitted at the Bay Front beach in Hamilton Harbour just 62 percent of the season.[52]

Myriad problems continue to beset the sewage system, resulting in frequent environmental contamination. The worst incident occurred in January 1996, when a pump failure at the Woodward Avenue treatment plant flooded more than 100 homes and businesses and spilled 182 million litres of raw sewage mixed with rain and melting snow into local creeks and Hamilton harbour. Factions disagreed over who was to blame. PUMC's Dr. Smith denied responsibility, saying, "It could be an act of God for which no one is to blame. It could be the design of the plant for which the region is to blame. . . . We feel it has nothing to do with the way we operate the plant."[53] When PUMC's defenders pointed to a gauge malfunction that misled the operator, its detractors countered that the operator should have known better than to switch off pumps during a heavy rain, regardless of readings on the gauges. A report prepared for the region by Filer Consultants charged that the incident resulted from the operator failing to perform appropriately in eight respects.[54] The Mini̇s of Environment also concluded that PUMC was at fault.[55] It did not, how ever, lay charges against the company, believing that its lack of evidence of adverse environmental effects would have prevented it from winning a court case.[56] Dr. Smith insisted that the spilled wastewater, extremely diluted, was cleaner than the receiving water and that "the bay was not affected one iota."[57] Regardless, as a "goodwill gesture," PUMC donated $27,000 to the Bay Area Restoration Council, a group monitoring the rehabilitation of Hamilton harbour.[58] It was harder to lay to rest the question of liability for flood damages. One-hundred-fifteen victims had claimed $2.5 million in damages, and squabbles with the region over who should pay dragged on for years. Not until 1999, when Azurix sought council's consent to PUMC's sale, did the region obtain relief: Azurix agreed to assume the liabilities associated with the spill.

A series of problems at the Woodward Avenue plant followed the 1996 spill. Levels of suspended solids and phosphorus periodically exceeded the baseline performance criteria set in the contract and the limits set by the province in certificates of approval.[59] PUMC reported that high water flows compromised effluent quality twice in 1996 and three times in 1997.[60] Local 772 of IUOE claims that the plant violated at least one of the conditions of its certificate of approval for four months in 1997.[61] Further violations occurred during the last three months of the following year; culprits included phosphorus, suspended solids, and biochemical oxygen demand.[62] May 1999 brought more compliance problems.[63] The plant also made the news in February 1999, when clogged screens caused sewage to overflow and run out of the building.[64] It again exceeded permitted levels of suspended solids and phosphorus in 2000.[65]

Even when the plant appears to be operating smoothly and is in compliance with provincial standards, it is often polluting. Dr. George Sorger, professor of biology at McMaster University, teaches students to monitor water quality. In the spring of 1996, students in his program found the effluent from the plant to be highly contaminated with total coliforms, E. coli, and ammonium ions and to have a very low level of dissolved oxygen. That summer, more than half of the samples of the water taken downstream from the plant's effluent pipe contained elevated levels of E. coli – in one case, 3,000 times the permissible level in recreational waters – and all contained elevated levels of ammonium ions. Students found the same situation in 1997 and again in 1998.[66] Although Dr. Sorger does not dispute that the plant is complying with standards, he challenges the adequacy of those standards: "In my opinion, the provincial standard for

...venue Sewage Treatment Plant is inadequate and we
... higher standard from the facility."[67]

years, another major concern at the Woodward Avenue plant has been the frequency with which sewage bypasses at least one stage of the full treatment process.[68] The inadequately sized plant reported more than 50 full bypasses and more than 300 primary bypasses in both 1997 and 1998.[69] In both years, approximately one-fifth of the wastewater entering the plant bypassed at least one stage of treatment.[70]

Other facilities have also experienced problems. Sewage spilled from a Waterdown pumping station in July 1998. September 1998 saw a six-hour sewage spill at the Waterdown treatment plant.[71] An electrical problem at the Dundas treatment plant caused sewage to back up and overflow through manhole covers in January 1999.[72] Sewage again flowed from a manhole in September 1999, when a Stoney Creek pump failed.[73]

Regardless, municipal staff generally – and many would say inexplicably – defend the system's environmental performance. In 1999, Mr. Gohier boasted, "The quality of the effluent is about 40 percent better than what is required under our certificate of approval. We're doing better than what's required by law and that's ultimately what counts."[74] Two months earlier, Mr. Gohier had acknowledged variances with the ministry's requirements but had maintained that they had been minimal and had posed no threat to the public or the environment.[75] Hamilton maintains that the contractor's environmental performance, however imperfect, is an improvement over its own previous performance. Mr. Gohier insisted in 1998 that "on average over the last four years on the wastewater side, the plant effluent under Philip has been superior to what it was when we operated the facility. That's a fact."[76] Chairman Cooke echoed that assessment the following year: "The benchmark for the contract is relative to where we would have been had we been managing the plant. You've heard they are either doing as well or better."[77] So, too, did Councillor Eisenberger: "They are meeting their objectives as good [sic] as or in some cases better than we did running the plant ourselves."[78]

An examination of effluent quality data confirms that the private operator's performance regarding biochemical oxygen demand has indeed been superior to that of the public operator. The data indicate that the Woodward Avenue plant exceeded the biochemical oxygen demand limit of 25.0 mg/L 11 times between 1990 and 1994 (when publicly operated) and one time between 1995 and 1999 (when privately operated); the daily averages were 16.8 mg/L under public operation and 12.0 mg/L under private operation. With regard to other contaminants, however, the private operator has fared less well. The plant exceeded the suspended solids limit of 25.0 mg/L two times between 1990 and 1994 and seven times between 1995 and 1999; the daily averages were 16.8 mg/L and 18.9 mg/L respectively. The plant exceeded the total phosphorus limit of 0.8 mg/L one time between 1990 and 1994 and two times between 1995 and 1999; the daily averages were .44 mg/L and .47 mg/L respectively. Levels of nitrogen and ammonia in the effluent also worsened over the decade; the daily averages for total kjeldahl nitrogen rose from 12.8 mg/L to 15.8 mg/L under public and private operation respectively, while the daily averages for ammonia rose from 10.6 mg/L to 12.9 mg/L.[79] When confronted with this evidence, Azurix president and CEO John Stokes admitted that the aging plant's effluent has, in some respects, worsened since 1995. He insisted, however, that the plant's performance has been better than it would have been had Hamilton not contracted out operations.[80]

Again, municipal officials leap to the operator's defence, pointing out that poorer effluent quality has often reflected circumstances beyond its control: The plant has treated stormwater and sewage that previously would have overflowed from combined sewers into receiving waters; the influent flowing into the plant has contained higher levels of ammonia and phosphorus; capacity limitations have prevented full treatment; and capital works in some parts of the plant have necessitated shutdowns that have caused bypasses or overloaded other parts of the plant.[81] In such cases, Hamilton cannot hold the operator responsible for its poor performance. The contract itself exempts the operator from responsibility for complying with environmental laws, regulations, policies, and standards under many circumstances: The firm is not responsible for non-compliance resulting from changes in the quality or quantity of the influent, contamination of the influent, limits of the facilities' capacity, or the region's failure to make capital expenditures.[82] Under such circumstances, the contract likewise protects the operator from failures to perform as well as Hamilton itself had been performing. Although the operator agreed in the contract to meet or beat the region's performance during the previous four years (measured by effluent quality), it secured a number of acceptable excuses for not doing so. It obtained permission to make "temporary excursions" from the performance criteria. Nor would it be responsible if poor performance were caused by an "unforseen circumstance" such as an extreme weather event, fire, sabotage, or an outside labour dispute.[83]

In a report prepared for CUPE, labour studies instructor John Anderson complained about PUMC's broad exemptions from liability, saying, "These clauses create an opening wide enough to sail the Titanic through."[84] The protections certainly seem to have muted criticism from bureaucrats and politicians. Mr. Gohier has consistently insisted that the fault for many of the environmental problems lies in the state of the sewage treatment facilities – which need $570 million in expansion and upgrades – rather than in the skill of its operators.[85] The problems, he told a reporter in 1998, are outside of PUMC's control. "That plant is an old plant, it requires a lot of capital works," he explained. "It's falling apart faster than we can fix it."[86] In defending the region's positive review of PUMC's performance throughout 1998, chairman Cooke likewise cast fault on the Woodward Avenue plant rather than on its operator: "We have a plant that is outdated, that needs improvement, that is struggling to meet present and future ministry standards."[87] Commenting on the same performance review, Councillor Eisenberger also refused to blame the operator for the plant's poor record: "Those are issues that would have happened, could have happened if the region was the operator as opposed to PUMC."[88] The following year, he attributed poor performance to Hamilton's long-term failure to invest the required capital: "The problems that the plant has experienced are problems of capacity and capital improvements that were identified 15 years ago and haven't been addressed."[89]

Even the environment ministry was slow to condemn the operation of Hamilton's wastewater system. Once a private firm assumed responsibility for operations, IUOE relentlessly pressured the ministry to enforce environmental laws, notifying it of violations, warning that the system could not be self-regulating, and complaining that the workers in the system were the only watchdogs.[90] But the ministry dragged its feet, preferring to review plans for compliance with the owner and operator rather than to lay charges forcing a clean-up.[91] In 1998, in explaining the ministry's refusal to launch an investigation into possible non-compliance at the Woodward Avenue plant, spokesman John Steele said, "We just can't, in all good conscience, say we're going to conduct an investigation when they have told us that they are working on the problem. It's an old plant, we know it has problems."[92] Regardless, a year later the ministry had begun to lose patience. Said Mr. Steele, "Obviously that plant has to show some improvement. If they don't, then we'll take the appropriate action, which means orders. They won't have any choice in the matter."[93] Why the owners or operators ever had a choice in the matter of compliance with environmental laws and regulations he did not say.

In the end, the ministry did act: Between June 2000 and January 2001, it laid 14 charges in connection with violations at the Woodward Avenue, Dundas, and Waterdown facilities in 1998 and 1999.[94] Another eight charges followed in July 2001, prompting a spokesman for the operator to wonder out loud if his company were being treated more harshly than its public counterparts with similar or worse operating records.[95] That certainly appears to be the case. At the very least, the union, the public, and the ministry are holding the private operator to higher standards than those to which they held its public predecessor. The ministry did let up its pressure somewhat: In June 2001, it granted relief from further charges for violations pending the completion of the improvements being made to the system, which were expected to be completed by August 2002.[96] Nonetheless, in August and November 2001, Azurix was fined a total of $217,600 as a result of the charges already laid.[97]

When privatization was first proposed, financial savings joined job creation and environmental benefits on Hamilton's list of anticipated gains.[98] Certainly, the guaranteed annual savings – comprising just 3 percent of the amount that Hamilton had previously budgeted to run the plant – are modest compared to the 25-to-50 percent savings achieved in jurisdictions that have contracted out operations through a competitive bidding process.[99] It has been difficult to determine the extent to which privatization has created savings beyond those guaranteed in the contract. For years, even the region itself did not know if, or by how much, savings exceeded the minimum. Although the contract specifies that the contractor's auditors will calculate cost savings within 90 days of the end of each contract year, in April 2000, the region and the contractor were still negotiating the figures for 1997, 1998, and 1999.[100] Azurix confirmed in December 2000 that Hamilton had not received any payments under the contract's profit-sharing formula.[101]

The difficulties in calculating savings reflected the challenges associated with distinguishing operating expenses from capital investments, and budgeted expenditures from unbudgeted expenditures. They also reflected uncertainty surrounding the base budget. Dr. Smith, describing the guesswork involved in determining Hamilton's cost of running its water and wastewater system before contracting out, said that it took one almost-full-time bureaucrat an entire year to tease out the costs.[102] The complexity of the rules governing which party should bear which costs doubtless also increased the difficulty of calculating the region's savings. Under the contract, the region assumed the risk of numerous increased costs. The base budget accounted for inflation on fixed costs, variation in the costs of hydro, natural gas, chemicals, and the disposal of ash, grit, and sludge, and increases in corporate overhead.[103] The contract permitted the operator to pass on to the region a wide variety of other costs as well, providing for the recalibration of the base budget under numerous scenarios causing spending to rise more than anticipated. These scenarios included: the issuance of court orders; changes in laws and regulations; increases in permit, licence, professional, and filing fees; alterations to disposal of ash, sludge, and grit; enhanced grounds maintenance; adjustments to PCB reporting and inspection services; adjustments to the operation of a supervisory control and data acquisition system; fixing of operational parameters for wetwells, outstations, and the high lift; development of a maintenance management program; the region's refusal to make capital improvements; changes in the quality or quantity of influent; provision of additional services; employment-related claims arising from before 1995; and unnamed factors external to the operator's efforts.[104] In addition to making the calculation of savings more difficult, some such provisions may have weakened the operator's incentives to perform efficiently and effectively and may have increased the costs of privatization.

By June 2001, it appeared that Hamilton was enjoying a host of savings. In a presentation to council, Azurix estimated that benefits over the life of the contract would total $31.2 million (in nominal dollars), including operating savings of $7.5 million and avoided capital expenditures of $23.7 million.[105] (The avoided expenditures likely included the Woodward Avenue pre-treatment facility, the closed incinerator, the computerized automation system, and centrifuges.) Two months later, in a joint submission to the Walkerton Inquiry, Hamilton and Azurix identified savings of $35 million resulting from direct investment, operations savings, and additional revenue from a new carbon facility. They predicted that efficiencies will generate long-term operations savings of $1.8 million a year, which Hamilton will benefit from when the contract expires.[106] The city was promised further financial gains by American Water Works, which, in its bid to take over the Azurix contract, agreed to resolve several outstanding issues at a cost to itself of $680,000 and to invest an additional $975,000 in plant upgrades.[107]

Earlier estimates of savings had been far more modest. In October 2000, Councillor Eisenberger identified $10 million in savings from reduced operating costs and the Azurix-financed plant upgrade.[108] Councillor Wilson, chairman of Hamilton's finance committee, expressed mild disappointment in the cost savings. Although he did not quantify them, he conceded, "We haven't saved as much money as we thought." Nonetheless, he praised the operator's cost consciousness. "We have definitely saved money," he insisted. "We have saved by cutting operation costs and capital budget. Overall, the benefits have been positive for both. . . . Private operators are cost conscious; probably more than a council would be."[109] Chairman Cooke likewise attributed savings to the fact that the deal provides incentives for both sides to operate efficiently.[110]

Staff reductions have generated many of the operating savings. PUMC's contract with the region required the firm to retain existing staff for 15 months. Once that limit had passed, it lost no time in paring numbers. By mid-April 1996, it had laid off 20 unionized workers and three managers.[111] The layoffs surprised and angered politicians, who complained that PUMC had not informed them of its plans. The union also reacted critically. Mr. Yemen objected, "We understand that it may be necessary to lay off some people, but we're of the opinion they've gone too far."[112] The staff were in for far worse news in the following years: By 2000, several rounds of layoffs and early retirements left just 47 unionized and management employees in the system – a system that had maintained 122 positions five years earlier.[113] Not surprisingly, such deep cuts – along with management's expectations that the remaining workers would "cross-train" and "multi-task" – harmed labour relations, transforming IUOE from a supporter to a harsh critic and reinforcing CUPE's opposition to the deal. In fact, union opposition to the operator's introduction of a training program to facilitate automation, equip workers to perform a wide variety of tasks, and ensure literacy and numeracy led to a 111-day strike in 1999. Labour relations improved after the strike. By the end of 2000, Mr. Hoath wrote of enjoying "positive and productive relations" and developing "co-operation, trust, and mutual respect." The following year, representatives of both the operator and the union remained optimistic about their working relationship.[114] One measure of the attitudinal change is the decline in grievances, which fell from 56 in 1999 to seven in 2000 to zero in the first six months of 2001.[115]

Changes in plant processes – including the shutting down of the Woodward Avenue sludge incinerator and the closing of the Woodward Avenue boiler room following the transfer of heat from Hamilton's solid waste incinerator – made possible extensive staff reductions.[116] Computerized automation resulted in further staff cuts. Computers likewise brought about other efficiencies. A computerized maintenance management system enabled PUMC to monitor, record, and evaluate every facet of operations and expenditures, to isolate needlessly expensive processes, and to boost productivity. It also allowed PUMC to track the use of equipment and to plan preventative maintenance after a given number of hours of use, bringing savings of 40 percent, on average, over the costs of maintaining equipment after it has broken down.[117] In its review of PUMC's performance during 1998, the region praised the firm for increasing the ratio of preventative maintenance to corrective maintenance from the pre-privatization rate of 20:80 to 60:40.[118] By 2001, the ratio had improved to 70:30 – a marked improvement, but still short of the industry standard of 80:20.[119] Other savings have resulted from changes in treatment processes. For example, Azurix's decision to replace sewage sludge presses with centrifuges is expected to save $350,000 a year.[120]

Savings – like so many other aspects of the Hamilton privatization – have brought controversy. In two cases, the region and PUMC each claimed that the savings were rightfully theirs. The first such dispute arose over the savings to be generated from the transfer of steam energy from the region's solid waste incinerator to the Woodward Avenue treatment plant. The parties disagreed over the origin of the idea, the sharing of the capital cost (expected to be between $4 million and $8 million), and the distribution of the $2.5 million to $3 million in annual savings. The threat of a costly court battle inspired a compromise whereby PUMC and the region split the investment and shared in the savings.[121] Conflict also arose over the allocation of $1 million in annual savings arising from the closing of the Woodward Avenue treatment plant's sludge incinerator.[122] PUMC claimed a right to 60 percent of the savings, as they were realized under its management, while some regional politicians objected that the company should not benefit from process changes that had been in the works long before privatization.[123]

In a confidential appraisal of PUMC's performance throughout its first year of operations, then regional contract coordinator Robert Crane complained about the firm's overarching concern for its bottom line. He described inadequacies in PUMC's maintenance of facilities and equipment and noted confrontations over which projects qualified for payment under the capital budget, the number of personnel funded in the base budget, and the setting of the base budget. "The relationship," he summed up, "has been consistently confrontational, difficult, tense, and frustrating. In our opinion, PUMC's focus on this contract has been changing from a cooperative spirit of business development and economic development to one of profitability only."[124]

Other critics charge that some of the operator's savings have cost

Hamilton dearly. Greg Hoath, business agent with IUOE's Local 772, has accused PUMC of reducing the monitoring of equipment from once a day to three times a week.[125] Although reluctant to blame equipment failures or other accidents on staff cuts, he has speculated that a larger staff would have caught problems earlier – a suggestion that Hamilton rejects.[126] In a report prepared for CUPE, Salim Loxley and University of Manitoba economics professor John Loxley maintained that "the workforce has been decimated to the point where environmental standards cannot be met." "Some observers," they noted, "feel that PUMC is choosing to operate at a much higher level of risk of environmental problems in order to reduce cost and increase its profits." Messrs. Loxley and Loxley suggested that because the contract exempts the operator from liability under so many circumstances, these increased environmental risks are borne by Hamilton.[127]

The structure of the contract may also give the operator incentives to offload more tangible costs onto Hamilton. Both Mr. Yemen and Mr. Hoath have charged PUMC with pumping too much water during off-peak hours when power rates were lower.[128] The additional pumping increased pressure in water pipes, straining them and causing some to rupture. Since PUMC was not responsible for the piping system, it had no incentives to avoid damaging it. (Further complicating the incentives was Philip's purchase of two water pipe repair companies – purchases that allowed it to profit from the ruptures.) Mr. Gohier initially countered that the growing number of water main breaks were part of a trend that began before PUMC took over operations of the system and were caused by the age of the pipes, a tenth of which are nearing the end of their life span.[129] But the region's growing concern about the problem led to a 1997 amendment to the operating agreement noting the suspected relationship between pumping pressures and water main breaks and specifying maximum pumping pressures.[130] Critics also maintain that the contract gives the operator incentives to stint on maintenance. Under the contract, the region remains responsible for capital costs and for maintenance costs over $10,000.[131] According to Mr. Hoath, this provision inspired PUMC to let the lighting in part of the drinking water plant deteriorate to such an extent that its repair could be claimed as a capital expense – a charge denied by Azurix.[132]

Such charges are difficult to prove or disprove. It is important to remember that the criticisms of the Hamilton contract have largely originated with IUOE – many of whose members lost their jobs because of the contract – or appear in work financed by CUPE, a union committed to discrediting privatization. None of this would matter if the criticisms were accurate. Both CUPE reports, however, appear to be far from accurate. Stan Spencer, then senior vice president with PUMC, called Mr. Anderson's report "misleading and factually incorrect."[133] Mr. Gohier likewise blasted Mr. Anderson's report, calling it "full of errors, half-truths and outrageously misleading statements loosely disguised as facts." The report, he charged, "as a result of being commissioned by CUPE, was basically written before the 'research' even started."[134] Even Mr. Hoath conceded that, in his haste to complete the report quickly, Mr. Anderson made mistakes: "There are a lot of things in there that are not factual."[135] The Loxley report, which contains at least three errors before the second sentence of the introduction, likewise tests credibility.

On the other hand, Hamilton has its own interest in promoting as a success its experience with privatization. The contract requires it to consider the operator's interest in all communications with the media on operational issues.[136] That provision should elicit considerable scepticism in interpreting municipal employees' remarks to the press. The contract also requires Hamilton, wherever possible, to support the firm's efforts to secure infrastructure contracts in other jurisdictions.[137] An early cooperative marketing agreement called for joint advertising in regional, domestic, and international publications and joint participation in environmental industry trade shows.[138] Other factors may have also tended to curb Hamilton's public criticisms of the operator over the years. As long as Philip remained involved in the contract, fear may have restrained some critics; the company had a reputation for being a bully and for vigorously pursuing those who criticized it.[139] Furthermore, an intricate web of personal and professional relationships between municipal employees and Philip or PUMC doubtless created numerous conflicts of interest. A number of municipal staff members have taken positions with PUMC and Azurix. Those moving through the revolving door include two managers – the region's senior environment director and its finance director – who helped negotiate the original contract.[140] In 2001, Mr. Gohier joined Azurix. Remaining municipal employees may be reluctant to criticize their former colleagues or may hope that favourable treatment earns them a position in the private sector. Municipal politicians and employees may also fear that too much criticism will doom the project, requiring a resumption of public operations. Those who remember the miserable confrontations over water and wastewater management in the past would be forgiven for being averse to running the system once again.

The absence of readily available information increases the difficulty of judging the positions of the different parties. Both Hamilton and its contractor have been extremely slow to disclose information on any aspect of the water and wastewater system. In one 16-week period of research for this chapter, an exchange of 28 letters, faxes, phone calls, and email requests with six different people at Azurix failed to produce even a single answer to questions about capital investment, cost savings, performance, labour relations, staffing levels, pensions, or public opinion. When Azurix officials finally did provide information (in a meeting and a series of emails and telephone calls), they identified several important details as confidential or "draft" and insisted they not be cited. Although most of the contract is publicly available, PUMC insisted that key portions of the public copy be excised to prevent the disclosure of proprietary information or information whose release would be injurious to its business operations. Some of the details of PUMC's compensation and economic development requirements are therefore safe from scrutiny. The contract further disempowers the public by not requiring the operator to provide information – including treatment results, reports on the status of repairs and maintenance, and audited financial statements – to the public.[141] Nor does it require the region to make public this information or its own performance reviews. In fact, the contract specifies that sampling and test results are the property of the region and may be confidential.[142] Messrs. Loxley and Loxley called the lack of accountability to the public "perhaps the most disturbing element" of the privatization.[143]

Most likely, both truth and "spin" characterize many statements on both sides of the debate over Hamilton's privatization. Both praise and criticism are warranted, since the deal has succeeded in some regards and failed in others. Such variance is well reflected in the wavering comments of Councillor Wilson, who chaired the region's environmental services committee when the deal was signed and later chaired Hamilton's finance and administration committee. In 1996, he sounded worn out by the region's on-going disputes with PUMC. "It's been a lot tougher slogging than we anticipated," he said. "We expected there would be a lot of hills and valleys. But there have turned out to be a lot more hills. . . . We've spent a disproportionate amount of staff time dealing with PUMC over fairly small items. Part of the problem is there wasn't enough time to properly negotiate the contract in the first place."[144] By 1999, Councillor Wilson sounded more comfortable with the deal. Despite his initial reservations, and despite his general scepticism about privatization, he said, his fears were not borne out. "I think we put in sufficient controls and we're in pretty good shape. . . . There have been a few glitches now and then, but we had glitches too."[145] Even so, he said the following year, "The bloom is off the romance a bit."[146] Similar ambivalence is reflected in the *Hamilton Spectator's* assessment of the contract. On one hand, an August 2001 editorial noted, the city has saved on operating and capital costs and the contractor is helping to modernize the main sewage treatment plant. On the other hand, "privatization has not been without its problems." The editors cited concerns about the absence of competition, limited public accountability, and ongoing sewage pollution. They did not, however, condemn privatization: "The city's experience with privatization of the water system is still a work in progress, with mixed results. We do not share [Councillor] Merulla's view that privatization is 'an experiment gone awry,' but there is room for improvement."[147]

Lessons Learned

In the first academic study of the Hamilton privatization, Nolan Bederman, working with University of Toronto law and economics professor Michael Trebilcock, expressed reservations about sole sourcing and emphasized the importance of competition. Sole sourcing, Mr. Bederman wrote, "fails to utilize an important element of the private sector. Sole sourcing fails to inject competition into the public project and hence, may often fail to achieve the efficiency gains that it set out to create. Because sole sourcing can also be the catch-phrase used to disguise nepotistic business ventures, its sincerity and commitment to the public good must be closely scrutinized." Mr. Bederman concluded, "it must be stated that for future public-private partnerships, especially those without special relationships and circumstances, governments should attempt to incorporate some elements of competition into their procurement process."[148]

Has Hamilton learned that lesson? Then-Councillor Geraldine Copps admitted, in hindsight, that the original contract should have been tendered.[149] However, council did not tender a new contract when Azurix purchased PUMC in 1999. Parties hotly debated the legality of tendering, with some insisting that the financial troubles of Philip Services gave council the right to cancel the initial contract and others countering that voiding the contract would be premature and could cost the region as much as $100 million.[150] A number of water companies let council know that they would like to take over. United Water expressed its interest in either managing the balance of the existing contract or bidding on a new contract.[151] Councillor Ross Powers reported that six different companies told him they would bid on a contract.[152] The *Hamilton Spectator* commented in an editorial, "Waiting in the wings could be companies offering substantially better terms for the taxpayers. How will be know if council won't reopen the bidding?"[153] A number of politicians agreed. Ted McMeekin, then mayor of Flamborough, believed that opening the contract for bidding would be "the safest, cleanest, most prudent course." Competition, he said, would "test the real value of the asset, to get the sharpest price." Even Councillor Eisenberger, who was on the negotiating team and defended the Azurix deal, said that if the region had the opportunity to invite tenders, "I'd love to take it."[154]

Ultimately, council decided by a narrow margin – 14 to 13 – to approve the contract's transfer to Azurix. Supporters believed that the region secured important benefits: PUMC/Azurix agreed to finance, design, build, and operate a $7.5-million upgrade to the Woodward Avenue treatment plant, to resolve outstanding claims resulting from the 1996 sewage spill, to maintain corporate headquarters in Hamilton, providing a base of operations for Azurix's contract business in North America, to increase the number of employees at the headquarters by at least 25 percent, and to relocate an engineering division to downtown Hamilton.[155] Councillors defended their decision to go ahead on the grounds that if they delayed they might lose the benefits promised by Azurix and that they might lose control of the process to a bankruptcy judge once Philip Services entered Companies' Creditors Arrangement Act proceedings.[156] Others remained highly critical. Mayoral candidate John Munro likely appealed to many when he said, "I think the council's procedures (in awarding contracts without tender) are really hurting the reputation of this community."[157]

Officials from both the region and Azurix agreed that when the contract expires in 2004, it will be opened up for bidding.[158] Councillor Eisenberger called the untendered agreement "a unique, stand-alone arrangement." "I don't think we've done it anywhere else," he said, "and I don't suspect we will in other cases. Tendering is the fair way to go in most cases, unless there are unusual circumstances. This happened to be one of those circumstances."[159] When opening up the contract for competitive bidding, Hamilton should initiate the process well in advance and set aside the money necessary to negotiate the best possible deal. While rushing negotiations and scrimping on consulting and legal fees may bring initial savings, these are likely to be offset by the resulting long-term costs.[160]

When council revisits the contract, it will have the opportunity to do far more than open up the operation of its water and wastewater systems to competitive bidding. A broad range of changes are needed to protect the region from accusations, such as that levelled by Mr. Anderson, that it has adopted "the worst possible kind of privatization."[161] Council may want to rethink its objectives, de-emphasizing economic development and focussing on the quality of the service provided. If it maintains its economic development objectives, it must ensure that it does not do so at the expense of other important goals.

Council should negotiate an agreement that strengthens the contractor's incentives to perform effectively. To ensure improved environmental performance, and to transfer the risk of failure to the contractor, council will have to shore up the agreement's performance provisions and close loopholes allowing poor performance. It would be wise to introduce financial incentives for performance improvements and strict financial penalties for failures. It would also be wise to give the contractor the means to achieve the goals set out in the contract – to give it responsibility not only for operating and maintaining the system but for improving it. Assigning responsibility for capital improvements to the contractor – a change that would likely require a longer contract period and a new approach to setting water rates – would deprive the contractor of the excuse that Hamilton's inadequate investments are to blame for operating failures. It would also eliminate incentives to delay routine maintenance in order to turn operating costs into capital costs.[162] To further internalize costs, council should expand the contract to include the entire water and wastewater system, including distribution and collection. Doing so would deprive the contractor of the excuse that stormwater collection practices affect its effluent quality. It would also would eliminate any temptation to offload the costs of plant operations onto another part of the system.

Of course, performance provisions are only as good as their enforcement. It is essential that Hamilton understand that it has an *obligation* to enforce its contract – that its priority must be protecting the public interest rather than making excuses for or otherwise cooperating with a poorly performing contractor. Increasing the financial distance between council and the water and wastewater system (by assigning responsibility for capital improvements to the contractor, as recommended above) would reduce conflicts of interest that now discourage council from enforcing strict environmental standards. Once council no longer had to foot the bill, it would be more likely to insist on compliance. Of course, the users of the water system would have to bear the costs of meeting standards. But councillors would doubtless find it more politically palatable to attribute water rate increases to a private firm than to take direct responsibility for the increases. To further reduce potential conflicts of interest, council should close the revolving door between Hamilton and its contractors. It should also forbid campaign contributions by contractors to candidates in municipal elections. Such contributions, Mr. Anderson noted, "cannot help but undercut the perception of the complete separation of a company seeking favours from candidates who can end up adjudicating on them."[163] Finally, requiring full public reporting and subjecting the contractor to freedom-of-information rules would enable the public to better monitor the contractor's performance and to pressure council and the environment ministry if they shirked their regulatory responsibilities.

When Halifax, Nova Scotia, began looking at private-sector options for sewage treatment, former Hamilton councillor John Gallagher warned, "If you're using Hamilton as a model, you're making a terrible mistake."[164] His warning should sound in the ears of his former colleagues as they consider future options for the operation and regulation of their water and sewage system. They have the opportunity to move toward a competitive, efficient, effective, and accountable system. The first time, they got several important elements wrong. Perhaps next time, wiser from experience, they will get them right.

PART III

MAKING PRIVATIZATION WORK

CHAPTER 8

TURNING LOSERS INTO WINNERS
Bringing Workers Onside

Labour issues are the single biggest impediment to the privatization of public services in North America, dominating the concerns of politicians, bureaucrats, and business alike. In 1998, the Canadian Council for Public-Private Partnerships, the Canadian Chamber of Commerce, and the Federation of Canadian Municipalities sponsored a survey of 148 municipal, provincial, and federal administrators and politicians and 63 private-sector players. The survey found that labour issues present formidable challenges to privatization. The public respondents cited collective agreements and resistance from unions or employee groups as the greatest barriers to successful partnerships. Almost one-third of the pubic respondents and almost two-thirds of the private respondents agreed that "it's unlikely that a government would pursue a partnership if it faced resistance from a strong union." Forty-one percent of municipal respondents agreed that "union opposition could scuttle a partnership."[1]

The situation is no different in the U.S. Over a decade ago, the Reason Foundation's Robert Poole called public employee unions "the only significant organized opposition" to privatization.[2] In a survey of public managers by the Reason Foundation, 85 percent of the respondents identified public employee unions as a main obstacle to privatization.[3] The U.S. Urban Water Council found that 75 percent of the mayors surveyed listed labour as the largest inhibitor of privatization.[4] Surveys by the International City and County Managers Association, Touche Ross, and the Mercer group flagged similar concerns.[5] Labour concerns are most pronounced in larger communities, where the work force is more likely to be unionized. Smaller communities, where unions are neither as present nor as powerful, are less inhibited by the issue. A report on municipal privatization in Illinois, based on a survey of 516 responding officials from predominantly small, non-unionized municipalities, concluded that "opposition from labor unions, department heads and line employees was not cited as often as one might think"; the issue did not appear among the four factors most often mentioned as obstacles to privatization.[6]

Concerns about labour are based on the assumption that privatization harms workers. In 1998, a Vector poll sponsored by the Canadian Union of Public Employees (CUPE) found that 75 percent of the Canadians polled believed that privatization would result in the loss of good jobs, 73 percent believed that working conditions would deteriorate, and 72 percent believed that pay would suffer.[7] The following year, interviews of 66 Canadian opinion leaders – again sponsored by CUPE – suggested that such issues troubled those in public life or in the media more than those in business: Over 50 percent of those in the public sector, and over 70 percent of those in the media, were concerned that jobs would be lost and working conditions would deteriorate under privatization, whereas just over 20 percent of those in the private sector were concerned.[8]

Labour unions promote fears about privatization's impacts on workers. CUPE warns that public-private partnerships save the public sector money principally through staff reductions and cuts to wages and benefits.[9] It elaborates: "Good jobs are being replaced with privatized or contracted out jobs that pay less, don't have union protection, offer fewer benefits and are often part-time, temporary or casual. . . . Privatizing a public service wreaks havoc in the workplace – beyond the inevitable cuts to wages and benefits. As jobs get cut, workloads increase for those who keep their jobs, stress and tension levels rise, overtime demands skyrocket, and workplace harassment and violence become more frequent. Workers taking on additional burdens often pay a physical price as work-related injuries rise."[10] South of the border, the American Federation of State, County, and Municipal Employees (AFSCME) shares CUPE's concerns. A 1999 poll of over 1,000 AFSCME members revealed that four in five believe that privatization poses a very serious threat to jobs.[11] The union raises the spectre of costly layoffs, low-wages, fewer benefits, dead-end jobs, and more social problems, warning that "adequate salaries and fringe benefits may be sacrificed to provide greater profit to company owners."[12] An ocean away, the Public Services International Research Unit (PSIRU) sounds a similar alarm, complaining that contractors reduce costs through cuts in jobs, pay, and working conditions.[13]

Such concerns explain labour unions' passionate opposition to privatization. Around the world, unions have launched well organized and generously funded campaigns featuring lobbying, public relations, demonstrations, court cases, and strikes. CUPE is leading the Canadian campaign against privatization. In 1997, the union's national convention adopted a comprehensive "action plan" to defend public services. Under that plan, the union launched a campaign to fight privatization and contracting out, choosing water privatization as one of its three national strategic targets. In 1999, it announced that it was stepping up its fight against privatization.[14] Its first major targets – one unsuccessful and one successful – were the private design, construction, and operation of several wastewater treatment plants in Halifax, Nova Scotia, and a water filtration plant in Vancouver, BC. The choice of targets was interesting, given that there were no union members with vested interests to protect at the as-yet unbuilt plants. CUPE's opposition suggests how nervous it is about the possibility of any successful privatization – even one that does not directly affect it – occurring in Canada.[15]

Labour unions' fears are understandable. Dramatic job losses have followed a number of water and wastewater privatizations, most remarkably in England and Wales, where privatization has cost almost 20,000 workers – 41 percent of the industry workforce – their jobs. Although the total numbers are lower elsewhere, the percentages are often as high or higher. At a conference hosted by the Canadian Council for Public-Private Partnerships, moderator Harry Swain suggested that the layoff of 60 percent of the workforce is often feasible.[16] The operator of Hamilton, Ontario's, water and wastewater system laid off more than 60 percent of the facilities' workers.[17] The 165 positions eliminated at Indianapolis's wastewater facilities comprised 51 percent of the pre-privatization staff. When Adelaide, Australia, outsourced water and wastewater services, 340 employees – 49 percent of the workforce – lost their jobs.[18] Considerable job losses likewise followed the 1998 privatization of Milwaukee's wastewater system. When the contract was signed, the Milwaukee Metropolitan Sewerage District had 325 positions budgeted, 300 of which were filled. The contractor expects to reduce staff to 191 within three years – a cut of 41 percent.[19]

Staff reductions of such magnitude obviously have a significant effect on labour unions, reducing their numbers and, ultimately, their power. However, they need not adversely affect individual workers. As the Reason Foundation's William Eggers explains, "One of the big secrets about privatization is that it's almost never as tragic for public employees as you hear."[20] It is important to distinguish between job losses and *involuntary* job losses. A number of studies have indicated that privatization does not result in the latter as often as is widely assumed. In 1989, the National Commission on Employment Policy – an arm of the U.S. Department of Labor – examined the privatization of 34 city and county services. Of the 2,213 government workers affected, only 7 percent were laid off. Fifty-eight percent obtained positions with the private contractors; 24 percent transferred to other government jobs; and 7 percent retired. "In the majority of cases," the study concluded, "cities and counties have done a commendable job of protecting the jobs of public employees."[21] A review of the privatizations in Los Angeles in the mid-1980s likewise uncovered few involuntary job losses: Of the 1,321 workers affected, just 36 – or 3 percent – were laid off.[22] The Illinois survey produced similar results: Only 3 percent of the responding municipalities reported layoffs, while 65 percent reported that privatization had no effect on municipal employees.[23]

To protect workers, many communities encourage contractors in a variety of industries to reduce staff only through attrition, early retirement, or outplacement. Such protections rarely necessitate maintaining high staffing levels for long. With attrition rates averaging 6 or 7 percent in the U.S., numbers can fall rapidly without layoffs.[24] Generous early retirement incentives can likewise reduce staffing levels quickly. The privatizers of British Airways reduced staff from 59,000 to 39,000 without firing anybody, primarily by offering liberal terms of voluntary early retirement.[25] Perhaps because they need not be onerous in the long term, "no layoff" policies are commonplace. Many jurisdictions provide for "successor rights" that transfer the staffing protections in collective agreements to new employers. Ontario's Labour Relations Act vests municipal employees – but not provincial employees – with successor rights.[26] Los Angeles County asks privatizers to rely on attrition or transfers to county government positions.[27] While New York State does not forbid lay-offs, it protects affected workers by placing them on a redeployment list and offering them jobs – maintaining comparable salaries and titles – elsewhere in the state government when they became available. All six jurisdictions studied by the U.S. General Accounting Office provide safety nets for displaced workers. In addition to reassigning workers to other government units, they offer early retirement packages, severance pay, buyouts, and job transition assistance such as career planning, education, and retraining.[28] This wide-spread commitment to protecting workers may explain why AFSCME's 1999 poll revealed that fewer members were worried about job security than were in a similar poll conducted two years earlier.[29]

The impetus to avoid layoffs can come from a private firm rather than a government. Firms increasingly understand that avoiding layoffs may be in their best long-term interest. In a sense, unions are their customers: They are capable of scuttling a deal with a firm that has a bad reputation. As Mr. Stitt explains, firms "want to be able to go into the next city and the next city and have that labour organization vouch for them having been treated well." It was this concern for reputation – and for winning future contracts – that prompted the firms involved in the privatization of Indianapolis's airport to refuse to lay off employees, even though the city had given them permission to do so.[30]

Many water and wastewater privatizations have incorporated these transitional strategies to protect workers. Contracts frequently forbid firms from laying off employees for a specified time period – six months in Clermont County, Ohio; 18 months Scranton, Pennsylvania; three years in Birmingham, Alabama; 10 years in Milwaukee; and 20 years in Atlanta, Georgia, and Springfield, Massachusetts.[31] Full-hire provisions are becoming the norm in long-term contracts.[32] Alternatively, cities or contractors often ensure that displaced workers are not left without jobs. The firm that won the contract for Stonington, Connecticut, offered all displaced workers positions at other facilities within its network.[33] In Indianapolis, the city and its wastewater contractor designed a "safety net" to protect workers not offered jobs. Workers could choose between a job with another city department – the city had banked vacancies in anticipation – or a severance package that included counselling and out-placement services funded – to the tune of US$300,000 – by the contractor. Sixty-seven workers remained with the city while 43 found jobs through the outplacement program.[34] The contractor also agreed, in the event of future layoffs, to make its best effort to place affected workers in comparable positions at other facilities operated by the company, its partners, or its parent companies.[35] The commonness of such agreements suggests that workers' fears of being left without jobs may be more imagined than real.

Fears of individual financial insecurity resulting from job losses may also be overblown, thanks to generous severance packages. According to the Amalgamated Engineering and Electrical Union, workers in the UK water industry enjoyed attractive redundancy, severance, and early retirement packages – packages that made job losses tolerable, if not welcome: "In many instances it was financially lucrative for employees to volunteer to leave the company. In some instances it was simply 'an offer you couldn't refuse.'"[36]

Unions' claims about privatization's effects on compensation may likewise be founded more on fear than fact. CUPE's charge that average hourly wages in the private sector are far lower than those in the public sector – $13.85 an hour compared to $19.38 an hour – appears meaningless, as there is no indication that the comparison comprises similar jobs.[37] Contrasting CUPE's findings are those, in the U.S., of the National Commission on Employment Policy, which pointed to higher wages in the private sector and better benefit packages in the public sector.[38] A

1993 study by the labour-supported Economic Policy Institute confirmed these findings. The study challenged several reports that had concluded that total compensation packages were higher for public-sector employees than for private-sector employees, saying that they had ignored the issue of comparability. The authors found that after adjusting for differences in education, experience, occupation, and several other factors, "local governments paid their employees 4.5 percent less than similar private sector workers." Although the authors noted the difficulties of comparing the values of widely-varying benefit packages, they reported several studies indicating that public sector workers have more generous health insurance and pension plans.[39] *Public Works Financing* agrees that while comparable pay is the norm, few private utilities match the generosity of public pension systems.

Whether necessary or not, many communities ensure that employees do not pay for privatization through cuts to salaries and benefits. Toronto's fair wage policy requires all bidders for city contracts to pay their workers at least 94 percent of union rates, even if the workers are not unionized.[40] More than 40 U.S. cities, inspired by Baltimore's example in 1994, have passed "living wage" laws requiring firms contracting with the city to pay wages considerably above those specified in federal minimum wage laws.[41] Baltimore's law requires private companies that violate the rules not only to make restitution to affected workers but also to pay the city a penalty. Repeat violators may be excluded from bidding on city contracts for up to three years.[42] Massachusetts has gone even further, passing an anti-privatization law known as the Pacheco Law in 1993. Among other things, the law requires contractors to compensate former state employees at a rate comparable to their government pay and benefits.[43] It also establishes a maximum salary for managers of private service providers, preventing them from earning more than the highest paid state managers.[44]

Many challenge the appropriateness of wage and benefit guarantees. Some maintain that if public workers are indeed paid more than those in the private sector, they should not be, since their gains are others' losses. The Reason Foundation notes the inequity of such arrangements: "Any excess compensation enjoyed by public-sector workers is gained at the expense of taxpayers. Rather than being a reason for not privatizing, any 'compensation gap' is evidence that public workers are enjoying a pay premium gained through political leverage. Introducing competition to public services helps ensure that citizens receive fair value for their tax dollars."[45] The Massachusetts governor's spokesman criticizes the

Pacheco Law for its failure to serve taxpayers: "Senator Pacheco's real agenda is not accountability to the taxpayer; it's accountability to the public employee unions."[46] Accountability to unions can prove expensive. The Independent Contractors Association of Ontario claims that Toronto's fair wage policy costs the city $200 million a year.[47] Others complain that wage and benefit guarantees reduce potential cost savings and discourage many firms from competing. Mr. Eggers calls living wage laws "a backdoor way to kill privatization that has proven to be effective so far." He levels the same criticism against the Pacheco Law, explaining that in the two years preceding the law, Massachusetts had privatized over 25 services, whereas in the three years following the law, it had not privatized any. The law, he notes, has been so effective that it has become "a model for unions fighting privatization."[48]

Regardless of its merits, wage and benefit protection is increasingly popular in water and wastewater privatizations. As privatization consultant Mr. Stitt explains, "That's almost become a non-starter if you're not going to do that in the U.S."[49] The Indianapolis wastewater agreement requires the contractor to provide employees with a package of compensation and benefits equivalent to or better than that provided by the city prior to privatization.[50] The resulting private salaries and pensions have been so generous that public sector employees now want to use them as the benchmark for negotiating their own compensation.[51] The Milwaukee contract likewise commits the operator to equalling or exceeding the compensation and benefits packages offered by the district, as do those in Gary and Atlanta.[52] The Milwaukee arrangement breaks new ground by allowing the former city sewage workers to remain in Milwaukee's public pension system. *Public Works Financing* predicts that this arrangement may pave the way for similar deals in other cities, easing labour's concerns about privatization.[53]

Asset sales may offer workers the opportunity the benefit financially in another way: through share purchases. The World Bank's analysis of 12 asset sales in Chile, Malaysia, Mexico, and the United Kingdom, which considered workers not only as wage earners but also as buyers of shares, concluded that none of the sample divestitures made workers as a whole worse off and that several of them made workers significantly better off.[54] The Adam Smith Institute's study of developing and post-communist countries also found that employees have benefited from privatization as shares they purchased have increased in value.[55] Privatizations in the U.K. have frequently incorporated special stock options for employees. In some cases, workers have been allowed to pay for their shares from future earnings. Other times, they have been able to purchase shares at discount prices. Such offers have proven extremely popular, with an average take-up of over 90 percent. And they have helped generate worker support for privatization. As Madsen Pirie, president of the Adam Smith Institute, explains, workers will generally choose their own enrichment over solidarity with their union leadership.[56]

But privatization need not drive a wedge between workers and their unions. In fact, privatizers are increasingly allowing private sector workers to remain members of public sector unions. This is critical to many unions, which suffer declining memberships and which see in privatization a threat of further decline. In Canada, where 71 percent of public sector employees – but just 18 percent of private sector employees – are unionized, their concerns are understandable.[57] Like other concerns, however, they may be overblown. Successor laws or clauses in contracts often bind new employers to current union contracts and oblige them to recognize and negotiate with the unions when the contracts expire. Experience south of the border confirms that privatization and unionization can be compatible. The Illinois survey found that in 42 percent of the municipalities that had unionized workers, the private firm recognized the union.[58] In Indianapolis, as noted earlier, the wastewater contractor signed a collective bargaining agreement with AFSCME. Milwaukee's wastewater contractor signed a unified collective bargaining agreement with all four of the system's union locals. The employees will revert to being public employees at the end of the contract. Predictably, AFSCME had fought the privatization. However, District Council 48 executive director Richard Abelson now concedes that if outsourcing is inevitable, "Obviously, this is the model we'd like to see elsewhere." He adds, "This has been as positive an experience as a privatization could be."[59]

Private firms may also provide workers with a safer environment. Wastewater privatization in Indianapolis brought with it a different attitude toward safety, described by long-term employee Tom Wolf as follows: "Safety here used to be an afterthought, but now it's an everyday consideration." The resulting 84 percent decline in workplace accidents elicits praise from local AFSCME president Steve Quick, who notes that "employees respect the safer environment."[60] Employees doubtless feel the same way about Milwaukee's wastewater systems, where workplace injuries fell from 125 in the year before privatization to 44 in each of the two years following privatization – a 65 percent reduction.[61]

And yet, even without the threat of major involuntary job losses, reductions in salaries or benefits, decreased union representation, or compromised working conditions, many unions continue to oppose privatization. Perhaps it is inevitable. Perhaps, all else being equal, they simply prefer the protection of a political environment over one in which they and those they represent are judged on their merits. Mr. Eggers sums up the problem: "Privatization is disliked by the unions because it takes wage negotiations from the political arena into the market arena."[62]

Of course, the politicization of negotiations works both ways, benefiting not just the workers but also the politicians who create unnecessary jobs for them. One US study identified the pursuit of political ends as a principle reason for the public provision of services. "Politicians," it concluded, "derive significant benefits from in-house provision of public services – such as political patronage, support from public employee unions, control of unemployment through public payrolls – and may lose these benefits as a result of privatization."[63] Privatization robs politicians of such political benefits. As one individual involved in city politics told the *Detroit News*, "The mayor would no longer have jobs to hand out, the unions would no longer have employees to represent, and politicians at all levels of government would lose access to easy votes. If you eliminate the jobs and the unions through privatization, the mayor loses the political machine that gets him re-elected."[64] Economist Herbert Grubel made a similar point over a decade ago: "What we're asking politicians to do . . . is throw away one of the main instruments by which they compete with each other for votes."[65] The de-politicization of staffing and wage decisions also threatens politicians' war chests, since unions often contribute generously to election campaigns.

Outside of Canada, politicians' receptiveness to unions' concerns seems to be diminishing somewhat as the problems are proving solvable and as the offsetting benefits of privatization become ever more apparent. Perhaps workers too will begin to see that privatization offers them hitherto unacknowledged benefits. Although the private sector generally provides less security, it often provides compensatory advantages, such as better training. USFilter estimates that it spends close to US$5 million a year on training – US$2 million of that on formal classroom training.[66] Other benefits include more interesting challenges, greater incentives for improvement, and greater rewards for excellence. Such benefits are being acknowledged by union leaders in Indianapolis. AFSCME's Mr. Quick praises workers' opportunities in the new competitive environment, explaining that "city workers are no longer asked to park their brains at the door when coming to work."[67] He also has kind words for the city's wastewater system, recognizing that "More individual opportunities for training and advancement exist under WREP's management structure."[68]

Here in Canada, some unions – although not generally those involved in providing water services – are beginning to embrace privatization. The shift was evident at a conference hosted by the Canadian Council for Public-Private Partnerships in November 2000.[69] Joseph Mancinelli of the Laborers' International Union of North America sang the praises of partnerships. He enthused that the construction, operation, and maintenance jobs that partnerships in Central and Eastern Canada have created have been "fruitful and beneficial" for union members. Investments in the projects have created "fantastic" returns for the union. Furthermore, the union has enjoyed a greater say in the directions projects take and how they are run. The partnerships, he concluded, have achieved "a win-win situation for all of us."[70] Mr. Mancinelli was not alone. John Murphy, former president of Ontario's Power Workers' Union (the union that would later run full-page newspaper advertisements urging the Ontario government to privatize the province's electricity transmission and distribution systems), stressed the importance of being practical rather than ideological. Most public-private partnerships that he has seen, he said, have made sense from a practical perspective. Tony Tennessy, representing a BC local of the International Union of Operating Engineers (IUOE), told the conference that self-interest leads his union to support public-private partnerships for infrastructure projects, which are often huge employment generators. Had his union brothers in Hamilton been at the conference, they might well have disagreed with his reasoning, but even they appear to be coming around to his conclusion. The union's Greg Hoath says of public-private partnerships, "the concept is certainly a good one."[71]

By now, few municipalities fearing union opposition to privatization want for advice on how to proceed. Industry gurus commonly urge potential privatizers to be prepared; to develop transition strategies in advance of privatization, drawing from the elements discussed above; to engage unions and workers in the process; and to be committed to treating workers fairly. Mr. Pirie goes further. Disarm the opposition, he advises, by identifying all possible objections to privatization and tailor-making policies to deal with each one of them in advance. Do what it takes to elicit the support of those affected: Give people greater advantages than they previously enjoyed and never cancel a benefit. Mr. Pirie is confident that following such advice will "make friends out of your enemies," smoothing the way for a popular privatization.[72]

Even those following such advice will, without doubt, find it difficult to win over unions. Many surveyed by the Canadian Council for Public-Private Partnerships viewed union support as an oxymoron and stressed the importance of managing union involvement rather than gaining support. However, their pessimism may be unfounded. On face value, at least, union goals seem achievable. CUPE demands that unions have the right to organize and to negotiate collective agreements; that agreements and successor rights be respected; that wages, benefits, and working conditions be protected; and that unions be consulted, informed, and involved in reforms affecting them.[73] PSIRU's demands are similar: It insists that private firms should recognize local trade unions and protect workers' jobs, wages, and conditions. Privatizers increasingly meet such demands and are slowly earning grudging – albeit still very limited – praise from labour. It is too early to concede defeat in the battle to win union support. And doing so would be unwise. As one municipal employee warned, "It won't work otherwise."[74]

CHAPTER 9

THERE'S NOTHING LIKE A HANGING
Creating Incentives Through Legal Liability

In August 2000, in an unprecedented decision, a U.S. District Court granted motions holding Robert and Natholyn Adcock personally liable for violating the Safe Drinking Water Act. The Adcocks, owners of 11 drinking water systems serving more than 20,000 Californians, had been accused of violating regulations regarding microbiological contaminants, lead, and copper and of falsifying lab reports to hide the violations. They had argued that any liability that did exist should attach not to themselves but to their corporation. The court replied that the U.S. Congress had clearly intended to impose liability on persons or entities directly responsible for violating the SDWA. "Nothing in the SDWA," it explained, "suggests that Congress intended persons directly responsible for violations to be shielded from liability because they were employed by or acting on behalf of the corporation which actually owned the water system." The court found that it had sufficient evidence to hold the Adcocks liable on the basis of their personal conduct. The ruling left the Adcocks facing fines of up to US$27,500 per day for each violation.[1]

American sewage polluters likewise face personal fines for their wrongdoings. In November 2000, South Bay Utilities and its president, Paul Paver, were sentenced for violating the Clean Water Act by discharging inadequately treated sewage into Florida's Dryman Bay. The violations cost Mr. Paver US$455,000 in fines and restitution and cost his firm almost US$1.3 million.[2] The following month, a plea agreement between another sewage polluter and the United States Environmental Protection Agency confirmed personal liability for unlawful discharges. The owners and operators of a Puerto Rican hotel pled guilty to discharging sewage into the Caribbean Sea. Concho Corporation, one of the operators, was fined $300,000. Arnold Benus, Concho's president and majority shareholder, was sentenced to three years probation and fined US$130,000.[3]

These suits bode well for the future of water and wastewater treatment in the United States. The owners and operators of facilities around the country are doubtless taking notice. More important, reminded of their own liability for the performance of their systems, many will likely take greater care. To paraphrase Samuel Johnson, there's nothing like a hanging to concentrate the mind.[4]

"Hangings," in the world of water and wastewater, differ somewhat under our public and private legal traditions. They may result from prosecutions under public law, in which case they generally take the form of fines for violating statutes or regulations. Alternatively, they may result from law suits brought under private tort law. Under this long established but now frequently weakened form of liability, a party that violates others' common-law rights – usually, by acting negligently or by creating a nuisance – may face an injunction or may be required to pay damages to its victims.

However, liability is not merely about punishment or compensation. Above all, it is about prevention. The threat of liability for poor performance creates powerful incentives to reduce risks by performing well. Indeed, incentives and deterrence are at the heart of liability. As law and economics professor Michael Trebilcock explained of tort liability, "law and economics scholars more or less universally view the purpose of damage awards as creating appropriate incentives to take precautions on the part of injurers to avoid accident costs to potential victims."[5] Conversely, in the words of Supreme Court of Canada Justice John Major, "tort law serves as a disincentive to risk-creating behaviour."[6] Under a liability regime, a utility – or an individual employee – does not avoid harm simply because some regulation, to be evaded if possible, requires it to do so. On the contrary, it avoids harm because doing so is in its financial interest. In minimizing risks, it minimizes potential costs. Liability, in other words, internalizes the risks and costs of water supply and sewage treatment. It ensures that utilities' owners and operators bear the consequences of their chosen methods of treatment, testing, and delivery. Assigning full legal liability for poor performance is thus an essential means of promoting accountability.[7]

Tragically, in Canada, the providers of water and wastewater services have rarely been held liable for poor performance. They have rarely been prosecuted under statutory law or sued under tort law, either by governments or by individuals.

Paralysed by Conflicts: Why Governments Do Not Hold Public Utilities Liable for Poor Performance

Canadian governments almost never hold the owners or operators of water or wastewater systems accountable for damage done to public health or the environment. Governments lay charges for only a tiny fraction of utilities' frequent violations of statutes and regulations. In British Columbia, for example, in 1997, almost one-third of the province's worst polluters were local governments. Although many of the offenders had been polluting for years, the province had prosecuted none the previous year.[8] Ontario has been similarly loath to prosecute municipal polluters. According to the Sierra Legal Defence Fund, although the provincial government identified 187 violations of water pollution rules by municipal sewage plants in 1996, it prosecuted only one facility.[9] Nor has the province prosecuted municipalities that have failed to comply with drinking water regulations. Before the Walkerton tragedy forced it to pay attention, the province virtually ignored repeated studies documenting problems at a majority of drinking water facilities. The Canadian Environmental Defence Fund charges the Ministry of Environment with maintaining a *"de facto* non-enforcement policy for municipalities spanning three decades."[10] In fact, the province has known nothing but: Its policy of non-enforcement dates back as far as the late nineteenth century.[11]

These failures of regulation are hardly surprising, since, in our largely public systems, conflicts of interest prevent government regulators from doing their jobs effectively. Most provinces provide generous capital grants for municipal facilities. Regulators understand that strict law enforcement would mandate expensive repairs and upgrades for which their governments would ultimately have to pay. Furthermore, in some cases, provincial governments also operate water and wastewater systems. Ontario operates 161 municipal water plants and 233 municipal sewage facilities through the Ontario Clean Water Agency. Regulators understand that in prosecuting these facilities, they would, in effect, be prosecuting themselves. Even when municipalities operate their plants, regulators appreciate that these "creatures" or "children" of the province can hardly be considered independent. In such circumstances, self interest argues for a cooperative rather than a confrontational approach – an approach that emphasizes education, guidance, encouragement, and assistance over investigation and prosecution.[12] Conflicting loyalties and objectives thus paralyse governments that own, operate, or finance the water and wastewater systems that they must also regulate.

Economist Friedrich Hayek asserted that governments always shield public monopolies:

> A state monopoly is always a state-protected monopoly – protected against both potential competition and effective criticism. . . . Where the power which ought to check and control monopoly becomes interested in sheltering and defending its appointees,

where for the government to remedy an abuse is to admit responsibility for it and where criticism of the actions of monopoly mean criticism of the government, there is little hope of monopoly becoming the servant of the community. . . . The probability is that wherever monopoly is really inevitable . . . a strong state control over private monopolies, if consistently pursued, offers a better chance of satisfactory results than state management.[13]

European experience with public water and sewage utilities confirms Hayek's observations. As discussed in Chapter Four, France has been very reluctant to sanction publicly owned utilities.

In England and Wales, the conflict of interest between utility and regulatory functions was one of the reasons for the privatization of water and wastewater utilities. This conflict was debated in England and Wales in the early 1970s on the occasion of the creation of 10 regional public water authorities that would be responsible not only for treating sewage but also for regulating sewage pollution. Critics warned that expecting one body to act as both utility and regulator was as unwise as asking one person to act as both poacher and gamekeeper. The issue re-surfaced in 1987 with the following comments by the Secretary of State for the Environment: "In our consideration of the future for the water industry, we have been increasingly concerned by the role of the water authorities as both poachers and gamekeepers in this field. They are responsible for controlling discharges from industry and agriculture; but they are responsible for sewage treatment, and are major dischargers in their own right." The proposed solution? To privatize the utility functions, and to establish an independent authority to regulate the discharges.[14] David Kinnersley, who served first as a chief executive for a water authority and later as a board member of the new regulatory agency, identified the separation of the operator from the regulator as the "most significant gain" of the British water privatization. He praised the new "clarity of purpose in the different agencies," saying, "This could be a framework in which water utility privatization comes to be seen as sustainable."[15]

Indeed, in England and Wales, privatization greatly enhanced the enforcement of environmental laws. As the chairman of the environmental regulatory agency that was established upon privatization noted, Britain's old pollution permit system had been "designed with a view to avoiding an embarrassing number of failures and an excessive number of prosecutions of public organisations."[16] Accordingly, prosecutions were rare. The 1989 privatization changed that. Under the new system, prosecutions for environmental offences became the norm. By 1996, there had been 250 successful prosecutions of water and sewage companies.[17] The number of prosecutions increased even as environmental compliance improved. Nor does the trend show signs of letting up. In 1999, the Environment Agency prosecuted Thames Water on eight separate occasions; it took Anglian Water and Southern Water to court six times, Dwr Cymru four times, and Northumbrian Water three times.[18] Although the fines resulting from these prosecutions have often been low, considerable progress has been made.

Privatization in Canada will likewise help resolve the conflicts that now paralyse governments and will enable them to more vigorously enforce compliance with laws and regulations. With privatization, compliance shifts from the political arena to the contractual arena. Well drafted contracts clearly define and assign responsibilities. They set out specific, enforceable performance criteria and provide accountability mechanisms, such as fines, termination, or other penalties, to ensure that they are met. Privatization does not only provide tools for enforcement; it also provides incentives. Municipalities find it easier to insist that private firms meet tough performance standards. Stuart Smith, then president of Philip Utilities Management Corporation and former provincial parliamentarian, explained one contributing factor:

> It is generally easier for municipal authorities to demand performance from an outside contractor than from their own managers and staff. If you go to your own manager and you say, "We're not meeting the quality," and he says, "Yes, but I asked you four times for some additional people and you didn't give them to me," it's kind of hard to get into a fight at that point. But if you have a private company, you simply say, "You didn't meet the standards; you're out." It's a whole lot easier to demand accountability when you have somebody in that role.[19]

Milwaukee's experience with wastewater privatization exemplifies such a dynamic. In the words of the executive director of the metropolitan sewerage district, prior to privatization, "in-house staff could always blame insufficient staffing or insufficient funding for failure to perform timely maintenance, but our contract with UW sets performance standards that must be met within a fixed price."[20] Enforcing standards is even easier when the contractor bears responsibility not only for operations and maintenance but also for capital expenditures. Contracts that assign to private firms full responsibility for capital improvements and offload to these firms at least some of the political costs of increasing water rates to finance such improvements provide municipal officials with greater incentives – or, at least, fewer disincentives – to compel compliance.

In testimony to the Walkerton Inquiry, former Ontario Environment Minister Norm Sterling confirmed that privatization would bring better enforcement of regulations. He explained, "it's easier for the government to actually regulate the private rather than the public in some ways because – because you don't have any . . . political infighting taking place . . . in terms of that regulation. You have a private sector operator over there and then . . . you're not going to get any interference from the local municipal politicians." Mr. Sterling was then asked, "So there's no potential conflict of interest and no potential reluctance to prosecute or be involved in enforcement, is that what you mean? . . . You would be more willing to deal with them as the situation dictates, rather than for political reasons?" And he answered, "That's right."[21]

Former regulator Mervin Daub has confirmed that, in actual practice, private utilities tend to be subject to greater social control than public utilities. Comparing regulation of private enterprises to direct control through public ownership, the Queen's University professor of business identified "less direct political interference with economic decision-making" as an advantage of the former. Drawing on his experience as a member of the Ontario Energy Board (OEB), which during his tenure regulated three private regional gas monopolies and reviewed the rates proposed by the single public electricity monopoly, Mr. Daub concluded that if electric utilities were private corporations regulated by an agency such as the OEB, they would likely be subject not only to less political interference but also to "a more focussed, integrated, and regular form of control through the OEB, and much tighter scrutiny of performance through the daily attention of capital markets to equity values."[22]

That accountability inheres in privatization is well-recognized in the United States – so much so that some experts consider it the fundamental reason for privatizing services. In a 1996 staff report for the Joint Economic Committee of the U.S. Congress, senior economist Jerry Ellig wrote, "Privatization is based on the principle that private ownership generates greater accountability than the political process."[23]

The Reason Foundation has consistently stressed accountability as a primary benefit of privatization. In 1988, Reason president Robert Poole wrote, "Part of writing a good contract is to define measures of performance and allocate some city staff function to actually keeping track of what those performance levels are . . . One of the surprising benefits of privatization has been that when a service is contracted out, it's often the first time that anyone has thought about quantitative performance measures."[24] Mr. Poole expanded on these ideas in a 1999 speech delivered to a Toronto audience, saying that the request-for-proposal process may lead to a more precise definition of what needs to be provided.[25] Privatization increases the accountability not only of service providers but also of governments. In clarifying objectives, partnerships help de-politicize governments' decisions about services. Bill Eggers, director of Reason's Privatization Center, has also noted the need to de-politicize: "You need to physically separate policy from service delivery. There are all sorts of conflicting objectives if you have the same agency doing regulation, providing services, and giving policy advice."[26]

A 1996 Reason Foundation policy study comparing investor-owned and government-owned water systems elaborated on privatization as an enforcement mechanism:

> [T]he historical record indicates that government-owned companies have been less likely to comply with environmental and health standards than the investor-owned sector in a whole range of policy areas. Government-owned water companies are more likely to use their political leverage to fight stringent standards on whatever service they provide. In addition, the regulating agency has a more difficult time forcing government-owned companies to adopt the costly policies necessary to meet their standards. While the government can tell investor-owned companies to cut their dividends or operate with less profit, government-owned companies often demand increased subsidization, and thus increased taxes, to support any improvements. Since it is politically unpopular to raise taxes, the politicians have been known to look the other way on enforcement issues.[27]

In a privatized system, the market itself provides an additional element of accountability that regulators are unable to provide. The consequences of non-compliance are fundamentally different in the private sector, both for individual actors and for firms. Not only may a poorly performing operator or manager lose his job, but his firm may also lose both profits and clients, and its shareholders may lose their shirts.[28] A private firm's financial health depends upon its reputation. According to the Conference Board of Canada, one recent study concluded that 40 percent of the average company's market value is based on non-financial assets, including reputation. The board pointed to another study showing that 89 percent of consumers base decisions on reputation. Reputation, it concluded, is a strategic asset. But it can also be a precarious one. In the board's words, "Reputations take time to build but can be destroyed in minutes."[29] For these reasons, the deterrent value of liability is especially pronounced in private firms operating in a competitive market. Azurix president and CEO John Stokes succinctly described the consequences of irresponsible behaviour for a private firm: "If you are negligent, you are history."[30] The threat of being put out of business does not similarly threaten municipalities or their public utility commissions.

As important as privatization is in creating accountability, alone it is not sufficient to ensure enforcement. While privatization will reduce the conflicts that impede enforcement, it will not eliminate them entirely. Governments have long coddled private polluters and are likely to continue to do so. Additional measures are required to inform citizens and to enable them to act when governments drag their feet. The public must have ready access to information about water and wastewater utilities, and those harmed by utilities must be empowered to take court action when governments fail to protect their health and their environment. Individuals who have been adversely affected will not be restrained by the conflicts of interest or policy considerations that paralyse governments. As Justice Allen Linden explains, "Public servants are reluctant to move. . . . Politics may enter into the picture. An aggrieved person labours under no such burdens."[31] And yet, private citizens currently seem no better than governments at holding utilities accountable. They almost never sue over the commonplace violations of their rights. This is not because they would not like to do so. Instead, it is because a number of factors shield utilities – especially, but not exclusively, those that are publicly owned and operated – from full liability.[32]

Barriers to Liability Under Private Law

Historically, the law shielded the Crown from liability for most of the harms it caused. The traditional common-law doctrine of Crown immunity was rooted in the divine right of kings and the belief that the king could do no wrong. In the last half-century, a number of statutes have limited this ancient doctrine, repealing many of the immunities it conferred. In Canada, where immunities vary among the provinces and remain somewhat in flux, several common features continue to grant broad protections to the Crown. Governments enjoy considerable immunity from statutes, being bound by them only when expressly stated or necessarily implied. They also enjoy special protections from remedies available under tort law. Canadians cannot obtain injunctions against the Crown. In Ontario, it is the *Proceedings Against the Crown Act* that forbids courts from granting injunctions against the Crown or its servants and from making orders for specific performance.[33] Furthermore, although no such legislation prevents courts from awarding significant punitive damages against the Crown, they may be reluctant to do so, knowing that taxpayers, rather than the wrong-doers themselves, will ultimately have to foot the bill.[34]

A number of judges and legal scholars have found the doctrine of Crown immunity troubling. The Supreme Court's Mr. Justice Dickson wrote in a 1983 decision that the doctrine "seems to conflict with basic notions of equality before the law. The more active government becomes in activities that had once been considered the preserve of private persons, the less easy it is to understand why the Crown need be, or ought to be, in a position different from the subject. This Court is not, however, entitled to question the basic concept of Crown immunity, for Parliament has unequivocally adopted the premise that the Crown is *prima facie* immune. The Court must give effect to the statutory direction that the Crown is not bound unless it is 'mentioned or referred to' in the enactment."[35] Others have argued that private tort law should be applied to governmental activities. In proposing a rational law of state liability for the modern industrial democracy, lawyer William Bishop suggested, "If the state is simply one more participant in a normal activity . . . it should be liable as if it were a private person."[36]

Regardless, courts do not treat governments as private persons. Most significantly, they grant them broad immunity from the consequences of policy making. Governments, they rule time and again, should not be held liable under tort law for the results of decisions based on social concerns, economic expediency, or political practicability. Such decisions, at the heart of the political process, are not to be interfered with by the judiciary. Of course, courts do not give governments complete freedom to create harm. They distinguish between harms resulting from policy decisions and those resulting from operational decisions, condoning the former but condemning the latter. Mr. Bishop described the distinction as one between "a legally authorized imposition of loss and unauthorized sloppiness in the course of doing the authorized task."[37]

The Supreme Court of Canada discussed the distinction between policy and operational decisions in 1984. Madam Justice Wilson noted that it is often for local authorities to decide what resources to devote to specific tasks. They have to strike a balance between efficiency and thrift, and "whether they get the right balance can only be decided through the ballot box and not in the courts."[38] The Supreme Court reviewed the distinction between policy and operational decisions in 1989. Mr. Justice Cory explained that the Crown "must be free to govern and make true policy decisions without becoming subject to tort liability as a result of those decisions. . . . True policy decisions should be exempt from tortious claims so that governments are not restricted in making decisions based upon social, political, or economic factors. However, the implementation of those decisions may well be subject to claims in tort." Quoting the Australian High Court, the judge elaborated, "budgetary allocations and the constraints which they entail in terms of allocation of resources cannot be made the subject of a duty of care. But it may be otherwise when the courts are called upon to apply a standard of care to action or inaction that is merely the product of administrative direction, expert or professional opinion, technical standards, or general standards of reasonableness."[39] Mr. Justice Cory repeated much of this thinking in a 1994 decision. The incident complained of, he wrote, resulted from "classic policy considerations of financial resources, personnel, and, as well, significant negotiations with government unions. It was truly a governmental decision involving social, political, and economic factors." The Crown, he concluded, could not be found liable.[40]

A case regarding sewer backups in Thunder Bay, Ontario, illustrates the distinction between harms associated with governing and those associated with supplying services. The trial judge found that the city's failure to separate its storm and sanitary sewers stemmed from a lack of funds and therefore qualified as a policy decision; the city would not be liable for the resulting damages. On the other hand, the city's failure to enforce its policy of disconnecting rainwater leaders from the sewer system was not a policy decision. The judge, explaining that "inaction for no reason cannot be a policy decision," found the city negligent on this count.[41]

Statutory protections go even further to protect municipal governments' utilities from liability under tort law. Ontario's *Municipal Act* shields municipalities, council members, and municipal employees and agents from common-law liability for poorly operating water and sewage systems by forbidding nuisance proceedings in connection with the escape of water or sewage from water or sewage works.[42] Manitoba's *Municipal Act* offers similar protections, shielding cities from liability for losses or damages resulting from sewer overflows caused by ice or snow obstructions or by excessive rainfall. Likewise, the *City of St. John's Act* exempts the city from liability for flooding from rainstorms, thaws, or water or sewer pipe breaks caused by uncontrollable factors.[43]

Even when governments or their employees are held liable for the harm they cause, the consequences for perpetrators may be minimal. Those responsible rarely suffer. Union rules and a culture of secure tenure within governments reduce the risk of dismissal. Ontario's *Municipal Act* protects municipal employees from other risks through both the provision of liability insurance and the payment of damages or costs awarded against them.[44] Furthermore, if damages are awarded against a city, taxpayers or ratepayers will generally bear the costs. This off-loading of financial responsibility reduces the deterrent value of liability. Indeed, some scholars believe that governments' monopolies on services and their powers to tax themselves out of financial binds largely desensitize them to the incentives otherwise associated with liability. According to law professor David Cohen, "while the government might appear to be acting like a private firm, there is no reason to think that it will respond to economic signals when it is not constrained by private capital, product, and labour markets. Thus, a primary idea behind tort law – that both actors bear the full cost of the activity that generated the loss – is inapt when the government is one of the actors."[45] Others counter that punitive damage awards preserve liability's deterrent value. Even otherwise cost-insensitive governments may be managed by politicians or departments that want to minimize the tax burden or maximize their own budgets by reducing the waste resulting from large damage awards. In Mr. Bishop's words, "There is always a margin somewhere, a margin that the law can play on."[46]

Although governments do enjoy many special privileges, they are not the sole beneficiaries of limits on liability. Albeit less frequently, both statute law and tort law also protect private operations and the people overseeing them. Ontario's *Proceedings Against the Crown Act* grants protections not just to the Crown but also to independent contractors employed by the Crown. Whether those protected from common-law liability under Ontario's *Municipal Act* include water and sewage contractors is unclear. Neither the *Municipal Act* nor the *Interpretations Act* defines "agent." Should the term include independent contractors employed by the Crown, liability for faulty operations would apply to fully privatized utilities but not to those whose operations are merely contracted out to the private sector.

In tort law, both public and private utilities can, under some circumstances, use the "defence of statutory authority" to elude responsibility for nuisances they create.[47] This defence indemnifies a party from liability if its activity has been authorized by statute and if the nuisance is an inevitable result of exercising that authority. In order to earn the defence, the party needs to prove that there were no alternative methods of carrying out the harmful work and that it was practically impossible to avoid creating a nuisance. Courts, in determining whether governments have indemnified particular activities, generally distinguish between permissive and mandatory statutes. Under the former, which maintain industries' discretion over operating methods and locations, industries are expected to act in conformity with private property rights and cannot claim the defence of statutory authority. Only when the harm is an inevitable consequence of a legislatively authorized activity does the defence apply. In mandating an activity or authorizing something to be done in a specific manner or location, the reasoning goes, the legislature sanctions all of its unavoidable consequences, including those that would have previously been forbidden.

In 1989, two judges of the Supreme Court of Canada questioned the wisdom of the inevitability test, and with it the value of the defence of statutory authority. Chief Justice Dickson concurred with Mr. Justice La Forest that inevitability itself should not excuse exemption from tort liability. The fact that an operation will inevitably damage some individuals does not explain why those individuals should be responsible for paying for that damage. "Arguments about inevitability," the judges agreed, "are essentially arguments about money. . . . '[I]nevitable' damage is often nothing but a hidden cost of running a given system." Their conclusion? "The costs of damage that is an inevitable consequence of the provision of services that benefit the public at large should be borne equally by all those who profit from the service." The judges added that requiring the body that provides a service to bear the costs of its operations could serve as a valuable deterrent: "[I]f the authority is to bear the costs of accidents . . . it may realize that it is more cost-effective to forestal [sic] their occurrence."[48] Unfortunately, the judges do not seem to have succeeded in weakening the almost universal respect that the defence of statutory authority has enjoyed for over 200 years.

While tort law allows the defence of statutory authority only when harm inevitably results from an activity mandated by statute, Ontario's legislators have made the defence more broadly available to the operators of sewage works in the province. The protections date back to 1956, when the provincial government passed *An Act to amend The Public Health Act*.[49] The new law was the government's response to two court cases regarding sewage pollution.[50] In both cases, the defendant municipalities had argued that they were operating under statutory authority. The courts rejected these arguments, explaining that regardless of whether the government had approved their sewage treatment plans, water pollution was neither an anticipated nor an inevitable result; the municipalities, albeit at great expense, could have installed larger or better plants. The courts therefore issued injunctions forbidding the municipalities from dumping raw sewage into local rivers. The decisions alarmed the government, which feared that they would set an expensive precedent. It knew that 65 other polluting municipalities would be vulnerable to similar injunctions, that many lacked the capital – or the credit to borrow money – to pay for necessary repairs, and that it would likely have to foot the bill for any upgrades required by the courts. The government took the easier route: It dissolved the injunctions against the two polluting communities. It then extended protection to other polluting communities by deeming any sewage project approved by the Department of Health to be operated by statutory authority. These protections still exist: They now reside in the *Ontario Water Resources Act*.[51] As long as the works are in compliance with the *Ontario Water Resources Act* and the *Environmental Protection Act*, they are deemed to be operated by statutory authority. They are thus effectively immune to tort law challenges.

The different liability exemptions discussed above have one thing in common: They all externalize the costs of poor performance. Ratepayers, taxpayers, or the affected public are asked to pick up the tab for harm wrought by the utility. In short, liability exemptions are subsidies. And they are particularly dangerous subsidies at that, since, in shifting the costs of poor performance, they increase the likelihood of harm occurring. Where safety and environmental measures cost money, operators must find it hard to justify large expenditures if their liability is limited. As economics professors Richard Stroup and Roger Meiners explained, "When accountability is not achieved by liability, the incentive to use resources efficiently and to avoid harm to others is weakened. This is especially true when the law weakens the relationship between a faulty decision and payment of resulting costs by the decision-maker. The result is a poorer economy."[52] Another result is increased risk to public health and the environment.[53]

Strengthening Liability Under Public and Private Law

Clarifying and strengthening the liability – under both public and private law – of those who own and operate water and wastewater facilities would, more than any other single change, promote compliance with provincial regulations. But it would also do more: It would promote responsible performance beyond that required by regulation. The regulators of water and wastewater utilities can only achieve so much. The best regulators, no matter how attentive, will never match the best managers with regard to their knowledge of system operations. They will never as completely understand the effects on performance of operational changes or new investments. They will never as fully grasp the costs and benefits of selected methods, or as precisely determine the most efficient use of resources.[54] Nor will regulators ever completely control the day-to-day operations of the systems they oversee. They will never be as well equipped to foresee and avoid specific risks. It is therefore necessary to go beyond writing rules and regulations telling utilities how to operate. It is crucial to establish incentives for utility owners and operators to perform well along with disincentives for them to perform poorly – regardless of the regulatory environment in which they operate. Assigning wide-ranging legal liability for poor performance establishes such incentives. As law and economics professors Michael Trebilcock and Ralph Winter said of a different industry, "managerial incentives do matter: Increasing the relative benefits faced by management from increased safety will improve safety in a way that cannot be duplicated by regulation."[55]

Liability should therefore not be limited to its role in ensuring compliance with statutory laws and regulations. It should also be restored under tort law. Provincial legislators should ensure that nothing in provincial acts legalizes utilities' nuisances, thus maintaining citizens' common-law rights to sue. Regulators should replace permits granting absolute power to pollute – permits that sanction not just the polluting activity but the necessary consequences of that activity – with those permitting activities on the condition that they do not violate others' rights. Such conditions were common in nineteenth-century England, where early sanitation statutes maintained common-law rights by specifying that they did not legalize nuisances or other unlawful acts.[56] They remain common in the United States, where many statutes include "savings clauses" which preserve plaintiffs' rights to bring tort actions against those who harm them. For example, the Federal Clean Water Act specifies that "Nothing in this section shall restrict any right which any person (or class of persons) may have under any statute or common law to seek enforcement of any effluent standard or limitation or to seek any other relief."[57]

In the absence of statutes overriding the common law, those harmed by utilities could once again turn to the courts for injunctions and damages.[58] Those harmed by sewage pollution have traditionally relied on three branches of tort law for relief. They have turned to trespass law to stop direct invasions of their property, nuisance law to stop interferences with the use or enjoyment of their property, and riparian law to stop – or prevent – the corruption of lakes and rivers adjoining their property. Such suits were commonplace in both Canada and the United States before regulations preempted tort liability.

A typical riparian lawsuit of the early twentieth century involved Cobourg, Ontario.[59] The town had constructed a small sewer that emptied into a local creek. The creek crossed the grounds of a hotel, the owners of which objected to the discharges and sued the town. In its defence, Cobourg argued that the pollution from its sewer was not serious, as it neither smelled nor was likely to produce disease. The judge who heard the case agreed that the creek water was not very offensive. That, however, was not the issue. Riparians, he explained, have "the right to the water in its natural condition." The town, in contrast, had "no right to pollute this stream in the smallest degree." Cobourg argued that an injunction against sewage pollution would create a hardship for it. The judge responded with a passionate defence of individual rights over collective rights. Quoting an earlier court decision, he wrote, "I know of no duty of the Court which it is more important to observe and no power of the Court which it is more important to enforce than its power of keeping public bodies within their rights. The moment public bodies exceed their rights, they do so to the injury and oppression of private individuals, and those persons are entitled to be protected from injury arising from the operations of public bodies."[60] Accordingly, the judge issued an injunction preventing Cobourg from discharging any sewage into the creek. If contemporary courts were as free to protect individuals from utilities' pollution, our environment would benefit enormously.

This is not to argue that private tort law should replace public regulation. The two are by no means mutually exclusive. They can – and should – work together. The suitability of one or the other varies with the circumstances. Some scholars argue that the parties with the most complete knowledge should take the lead.[61] In determining health standards for drinking water, for example, a public regulatory agency will generally be more efficient than private parties. It is likely to have better access to – and be better able to evaluate – relevant medical knowledge. The same will often be true of sewage pollution. Private parties may be ill-equipped to determine risks or to counter them, especially if harms are slow to appear, widely dispersed, or difficult to attribute to a source.

On the other hand, regulators will not always have superior knowledge. Private parties will sometimes be better placed to evaluate the effects of sewage pollution. They will know of unique features in their environments that require unusual care. They will know how the pollution interferes with their activities and what the resulting costs are. They will know, in short, whether regulation is adequate. Indeed, as economist Daniel Benjamin has pointed out, they are ultimately the only parties with this knowledge: "only the people who bear the consequences of decisions can fully know the advantages and disadvantages of each expert decision . . . Ultimately, the uncertainty inherent in the process of risk assessment reinforces rather than obviates the need for individual resource owners to be the final arbiters of the risk assessment and decision-making process. There is simply no other way to choose (or choose to ignore) the experts, nor any way to weigh their findings."[62] When private parties find the regulatory system wanting, unimpeded access to tort law will protect their interests. Law and economics professor Steven Shavell explained the importance of supplementing regulation with tort liability as follows: "if compliance with regulation were to protect parties from liability, then none would do more than to meet the regulatory requirements. Yet since these requirements will be based on less than perfect knowledge of parties' situations, there will clearly be some parties who ought to do more than meet the requirements . . . As liability will induce many of these parties to take beneficial precautions beyond the required ones, its use as a supplement to regulation will be advantageous."[63]

One advantage of supplementing statutory and contractual liability with tort liability is that, while municipal governments may cap liability limits in contracts, they cannot limit liability to third parties under tort law. Many municipalities have no wish to cap liability. Indeed, the desire to shift risk to another party often contributes to decisions to privatize. But municipalities understand that assigning unlimited liability to the private sector has costs as well as benefits: It increases the costs of contracts and limits the field of competitors. The bid documents developed for Seattle's water plant procurements required full liability by contract or common law for third-party damages from operator negligence. The liability requirement dissuaded two companies from submitting final offers.[64] For its proposed Seymour filtration plant (before it chose the public route), Vancouver likewise insisted that proponents impose no limitations on liabilities for errors or omissions on their part. One short-listed company expressed reservations about the requirement. (Interestingly, one would-be competitor has expressed its willingness to accept unlimited liability in contracts for sewage plant operations but not in those for water plant operations, given the greater risks associated with the latter.) Some municipalities may reasonably decide that the costs of unlimited liability outweigh the benefits and may include liability caps in their contracts. In such cases, tort liability is an invaluable mechanism to ensure that individuals remain fully protected.

Tort liability has other advantages, as well. Unlike regulation, it compensates those who have been harmed. Although regulatory fines punish violators, they rarely help victims. Thus, justice is more fully served under tort liability. In many circumstances, justice is also more likely to be pursued. Private parties are often better placed than regulators to know when harms have occurred. Since they are directly affected, they often have greater incentives to stop the harms. The promise of a damage award strengthens their incentives to launch cases under tort law rather than statutory law. Furthermore, the standard of proof is lower under the former than the latter. In a tort action, the standard of proof is "the balance of probabilities" while in a statutory prosecution, it is "beyond a reasonable doubt."[65] For these reasons, private parties may be more likely than regulators to challenge polluters, and they may be more likely to launch private law suits than statutory prosecutions. By restoring full tort liability for the owners and operators of water and wastewater utilities, we can tap into a vast regulatory resource: the public.

CHAPTER 10

UNNATURAL MONOPOLIES
Lessons on Competition and Economic Regulation from England and Wales

In the best of all possible worlds, consumers would choose their water from a number of suppliers offering different qualities, services, and prices. Industrial consumers could reduce their costs by choosing inter-ruptible supplies or less treated water. Residential consumers could choose untreated water for their gardens and, for their personal use, could satisfy their preferences regarding taste and health concerns. The choice between fluoridated and unfluoridated water, or between chlorinated water and that disinfected by other means, would be made by individual households – or, at least, neighbourhoods – rather than city officials.

Such a world has been deemed hopelessly unrealistic by generations of water planners and economists who have taken it for granted that water is a natural monopoly that leaves little room for competition or consumer choice. The high cost of laying supply pipes, they have noted, makes it uneconomic for would-be competitors to install rival networks.[1] Such thinking, while rarely challenged directly, seems increasingly irrelevant. Economists are beginning to argue that although a true monopoly lies in transmission and distribution networks, water itself can – and, in some cir-cumstances, should – be supplied competitively. Michael Klein suggests that competition is warranted where water is scare and where tradeable rights to multiple sources exist.[2] Nicolas Spulber and Asghar Sabbaghi go further: They advocate multiple-pipe distribution systems delivering waters of different qualities produced by competing firms. They suggest that dual systems, which for new installations might cost 20 percent more than conventional systems, could be economic not for individual house-holds but for high-rise developments, for growth around urban periph-eries where distribution systems for low-quality irrigation water are already in place, or for areas where high-quality supplies are limited.[3]

Competition sceptics and others stuck in the natural monopoly rut need only look to England and Wales to see the error of their ways. Recent experimentation there points to myriad opportunities for competition in a privatized water world. Most consumers purchase water from one of the 26 regulated companies that, in their respective areas, are licensed to ser-vice all of England and Wales. But consumers may choose to purchase water from unlicensed private suppliers that meet quality standards but are not subject to economic regulation. Consumers who are willing to pay the costs of connection may also purchase water from regulated suppliers that do not normally service their area. Every regulated company has a duty to honour such "cross-border supply" requests from domestic customers and the choice of honouring or refusing such requests from industrial customers.[4]

To enhance competition, England and Wales have also pioneered "inset appointments" – the appointments of new licensees to service areas covered by existing regulated companies – for two categories of consumers: greenfield sites of any size and existing sites that use more than 100 megalitres a year. The new licensees need not provide their own sources of water or their own treatment facilities; they may use the existing companies' assets. In order to attract customers, they simply need to be able to provide better service or a better price. As many as 2,000 hospitals, airports, universities, breweries, manufacturers, and other large water users may be eligible for inset appointments.[5]

Ofwat, the economic regulator of the water industry, approved the first inset appointment in May 1997, when it licensed an existing regulated company to supply a large user previously supplied by another regulated company. Within three years it had granted eight inset appointments – five for greenfield sites and three for existing large users – and was considering a dozen more.[6] Ofwat broke new ground in May 1999 by granting an inset appointment to a new company – the first new licensed water company to be established since privatization. The new company obtained a license to provide services to Wales's largest water user, a pulp-and-paper mill. Although it intended to initially purchase water from the existing regional supplier and resell it to the mill, it planned to ultimately develop less costly water supplies, either by treating and reusing the mill's effluent or by tapping a new local source and reducing the degree of treatment. As the company's managing director pointed out, "you certainly don't need drinking water to make newsprint."[7] Inset competition facilitates such tailoring of services to meet customers' needs. As a result, it makes possible considerable cost savings: Inset appointments can save a typical user as much as one-third off the standard tariff.[8]

More radical than inset competition is England's nascent experimentation with common carriage, or the shared use of assets. Under common carriage, a water company (either a licensed provider or an unlicensed new entrant) may use a competitor's network of pipes to supply water to a customer. The Competition Act of 1998, which came into force in

March 2000, opened the door to common carriage by prohibiting water companies from abusing their dominant positions in the market. If an incumbent supplier's refusal to give a potential competitor access to its network is deemed an abuse of its dominant position, it may be fined up to 10 percent of its turnover or sued for damages.[9] In the fall of 2000, the regulated water companies published their "access codes" for common carriage. They laid out rules for setting prices for the connection to and use of their facilities and developed policies governing such issues as operating conditions, requirements for sampling and monitoring water quality, leakage allowances, emergency management procedures, and dispute resolution mechanisms. As no common carriage arrangements are yet in place, it is too soon to judge the codes' workability or their effect on competition.[10]

Both the government and the regulator continue to push the envelope. They are considering new models for competition, among them one that would separate the roles of retailer, network operator, and water producer and establish a market operator to set prices. In the mean time, Ofwat encourages competition where it can, be it in the bidding for contracts to lay mains for new developments or the trading of abstraction licences.[11] It does not admit to being discouraged. On the contrary, it insists that although progress towards competition has been slow, the mere threat of competition has brought pricing benefits. Water companies, more sensitive to their customers' needs, have honed their tariffs to better reflect costs and have introduced seasonal and interruptible variations.[12]

While competition in the product market remains limited, competition in the capital market flourishes. In competing for shareholders and their capital, water companies are forced to improve their performances and increase their profits. The threat of takeover inspires innovation and efficiency. Poorly performing companies cannot survive for long.[13]

Another form of competition infuses the regulatory process. Dubbed "yardstick competition" or "comparative competition," it involves the comparison of the British and Welsh water companies to one another and to their counterparts in other countries. The economic regulator monitors water company performance in order to determine the industry's best practices regarding service and price and to establish targets for individual companies. The regulator describes comparative competition as its main vehicle for improving company performance. Comparative competition, it suggests, "stimulates the behaviour that market competition might produce."[14]

Even in England and Wales, where new possibilities for competition continue to reveal themselves, the system remains essentially monopolistic for smaller water users. Domestic consumers cannot yet choose their suppliers. Even those who select an alternate supplier under common carriage will be unable to influence the quality of water that comes out of their taps, since water from different sources and treatment facilities will be mixed in the delivery pipes. (Although, of course, the ready availability of bottled water makes possible the choice of specific sources and kinds of treatment.) Most consumers will remain captives of their water companies. For them, water supply will remain, if not a natural monopoly, at least a *de facto* one. Strict regulation of water quality, service, and price is therefore essential to protect consumers from monopoly abuses.

In England and Wales, quality regulation is fairly straightforward: The Drinking Water Inspectorate oversees the quality of drinking water and the Environment Agency, relying on standards set by the European Union, regulates the water and sewage companies' impacts on the environment. It is in the field of economic regulation that England and Wales have experimented radically, with mixed results.

When the government began considering privatizing its water and sewage utilities in the mid-1980s, it called upon Stephen Littlechild for advice on economic regulation. Professor Littlechild – the economist who had helped develop the regulatory regime for British Telecom and who would later oversee the economic regulation of the electricity sector – recommended a departure from the rate-of-return model used to regulate private water utilities in the United States. Rate-of-return regulation controls profits by allowing a specified return on capital. Companies, confident that if they spend money they will earn money, have incentives to invest in their systems – an incentive not be sneered at in jurisdictions that need large amounts of capital. However, as Professor Littlechild pointed out, companies also have incentives to pad their expenses and to "gold-plate" their facilities. Furthermore, they have disincentives to cut costs and seek efficiencies, since they will have to pass any cost savings along to their customers.[15] And because the regulator must be highly involved in the process, reviewing expenses and investment programs, rate-of-return regulation can be cumbersome and costly.[16]

Professor Littlechild proposed a new system of regulation – one that would cap prices rather than profits and encourage companies to reduce costs through increased efficiencies. Borrowing the formula established for British Telecom, he proposed an "RPI-X" constraint on prices. Under this system, a company could not increase the weighted average of its prices by more than the rate of increase in the retail price index (RPI) less X percent. X, either a positive or negative number, would reflect both the investments required to meet environmental and health standards and potential efficiencies in operations. The regulator, using an industry yardstick to determine the scope for efficiencies, would review X every five or 10 years. Because companies could keep the money saved from increased efficiencies between the periodic reviews of X, they would have incentives to reduce their costs. The promise of profit, in other words, would drive companies toward efficiency. But the profit would be temporary: At every periodic review, the cost savings would lead to a tighter X and would be passed along to consumers in the form of lower prices. Profit-seeking companies would then have to become still more efficient, continuing a virtuous circle that would benefit themselves in the short run and their customers in the long run.[17]

Many years later, Professor Littlechild elaborated on the benefits of price-cap (or incentive) regulation, explaining that it "seeks to promote the kind of efficiency improvements that might be more associated with competitive industries than with regulated monopolies." Regulators, he continued, have neither the knowledge nor the incentives to mandate efficiencies; opportunities for cutting costs must be discovered by the companies themselves:

> Israel Kirzner has argued that even if one imagines a regulatory official dedicated to ensuring the adoption of all known possible measures for cutting costs, "one can hardly imagine him discovering, except by the sheerest accident, those opportunities for increasing efficiency of which he is completely unaware. The official is not subject to the entrepreneurial profit incentive. Nothing within the regulatory process seems able to simulate, even remotely well, the discovery process that is so integral to the unregulated market." The strength of the RPI-X approach is that it harnesses the incentives of the management within the market context, and does not require the official to know or discover the opportunities in question.[18]

The government accepted the bulk of Professor Littlechild's recommendations, substituting RPI+K for RPI-X in an acknowledgement that, because of the substantial capital investments required, prices would rise rather than fall. It set the initial Ks at an average of 6 percent – allowing prices to increase by 6 percent over inflation – and established Ofwat to set new Ks five years later.[19] In the interim, it gave companies permission to apply for adjustments in their price caps if the Environment Agency or Drinking Water Inspectorate imposed unanticipated costs on them.[20]

Price-cap regulation brought both expected and unexpected results. Companies promised to invest considerable capital in their systems, thereby justifying large price increases – overly large, in fact, when capital costs unexpectedly fell.[21] At the same time, they reaped profits by achieving remarkable efficiencies in their operations. As explained by the *Financial Times*, the scope for efficiency gains had been "grossly underestimated."[22] Shareholders therefore made far larger gains than anticipated – in some cases, as high as 30 percent a year.[23] The combination of soaring prices and profits created irresistible pressure on both the government and Ofwat to rein in the industry. The former reacted with a 1997 "windfall tax" clawing back £1.65 billion in past profits.[24] The latter further disciplined water companies at the 1999 review, when it imposed an average 12 percent reduction in prices.

Several water and sewage companies charged that these cutbacks were unsustainable. Share prices fell and investors left, making it more difficult to attract the funds necessary to invest in infrastructure.[25] The first company to balk at the 1999 price review was Kelda, the parent of Yorkshire Water. Kelda's board of directors complained that it could no longer make any money from owning a water business.[26] It proposed to sell Yorkshire Water's assets to a debt-financed, community-owned, not-for-profit mutual company and to obtain a contract to operate the new company's facilities. As a mere contractor, it anticipated, it would not be regulated and could earn higher returns. Although Ofwat opposed that plan, it did not rule out other attempts at restructuring.[27] Indeed, the following year, the regulator decided to permit, subject to several conditions, the acquisition of Welsh Water by a non-profit organization established to own the company and to contract out all operations and customer services to third parties.[28]

It is not yet clear how the industry will evolve, how the regulator will adapt, or how the companies' struggles with the regulator will be resolved. What is clear is that the regulator has to achieve a fine balance between the interests of producers and those of consumers. At the time of privatization, the government determined that Ofwat's primary duty must be to protect the viability of the water and sewage companies – to ensure that they would properly carry out their functions and that they would have the financial means to do so. Customer protection was deemed to be the regulator's secondary duty. The two duties are, in reality, more complementary than conflicting, since customers rely on a sustainable industry. As Ofwat has pointed out, "customers benefit if efficient companies remain financially viable."[29] Nevertheless, in response to the high prices and generous profits of the 1990s – and under the direction of the new Labour government – Ofwat became more sensitive to affordability and put greater emphasis on price reduction. It acknowledged that its efforts to reduce prices had to be tempered by the need to ensure that efficient companies had enough income to finance their functions. The regulator explained its challenge as follows: "Lenders and shareholders should be able to receive a return that is sufficient, but no more than sufficient, to induce them to make loans and hold shares, if the company operates efficiently."[30] Recent unrest in the industry suggests that Ofwat has not yet identified that just-sufficient return.[31]

Emerging interest in England and Wales in not-for-profit or community-held water companies that contract out operations to unregulated firms demonstrates that regulation, if mishandled, will encourage the industry to evade it. Experience in the United States confirms this. A 1995 study by the National Regulatory Research Institute (NRRI) identified regulation as a significant barrier to full privatization. It warned that "many privatization agreements are designed explicitly to avoid state oversight because economic regulation is perceived as overly bureaucratic and a threat to profitability."[32] Two of study's authors, Richard Dreese and Janice Beecher, later expanded on parties' reasons for wanting to avoid regulation:

> The desire to avoid state commission regulation of prices and profits is a recurring theme in privatization cases. Nonutility contractors and municipalities tend to oppose regulation, but for different reasons. Contractors do not want to be inhibited by regulatory procedures; more importantly, they do not want the state to limit their ability to earn profits. City officials generally are wary of regulation because they do not want to transfer rate-setting authority to the state. Cities also want to retain control over service boundaries and other franchise considerations because the management of local wastewater and water services is a key part of annexation strategies.[33]

Stuart Smith, then president of Philip Utilities Management Corporation, Canada's largest private water and wastewater management firm, expressed his own reservations about regulation to Ontario's Standing Committee on Resources Development in 1997. "We don't want to buy these utilities," he told the committee. "We don't want a regulating authority we've got to deal with."[34]

Neither consumers nor municipalities – nor, many would argue, even water companies themselves – will benefit from this apparent recoiling from economic regulation. In the United States, the now common rate-of-return regulatory framework was devised in the Progressive Era as a way to protect utilities from Populist governments' tendencies to fix prices at unsustainably low levels.[35] Utilities – and their customers – must continue to protect themselves against short-sighted politicians who rob water systems of revenues and capital in order to keep prices low or who use utilities for other political purposes, such as job creation or economic development. Good regulation assures all parties that legitimate revenue requirements will be met, that utilities will be able to attract capital, and that they will function efficiently. It thus has a stabilizing effect on utility finances and on the quality of services provided.[36] The authors of the NRRI study suggested that independent regulatory bodies may be best positioned to provide this kind of oversight: "State regulation generally is less parochial, less political, and less driven by expedience than local regulation. Without significant safeguards, local contracting and oversight can be prone to corruptive influences. It may also be somewhat easier for the states to make politically unpopular decisions."[37]

Independent bodies dedicated to the economic regulation of utilities also bring to their tasks greater resources and expertise than most municipalities can muster. They are better equipped to deal with the complex issues surrounding revenue requirements (including rate of return, economic efficiency, accounting policies, and operating and capital budgets), cost allocation, and rate structures. Some industry analysts argue that regulators' superior expertise positions them not only to control privately owned utilities but also to oversee the pricing of water and sewage services delivered by publicly owned utilities and the letting of operations and maintenance contracts (which, in controlling prices and enshrining standards, are themselves regulatory instruments). Such oversight is rare in Canada, with the notable exception of Manitoba, where the Public Utilities Board regulates the rates charged by water utilities outside of Winnipeg.[38] Nor is it common in the United States, where 46 states have established commissions to regulate privately owned water utilities, but just 12 state commissions have any authority over publicly owned systems, and only a few of these oversee operations and maintenance contract arrangements.[39]

Noting that long-term franchise bidding can be extremely complex, that if mishandled it can reduce rewards for capital investment and create incentives to mine a utility's capital at the end of a contract period, and that "the success or failure of franchise bidding rests on the strength of the underlying contracts," utility analyst Thomas Adams advocated subjecting contracts to one-time regulatory oversight.[40] Ms. Beecher spelled out a number of other reasons for regulating contracts. Regulation, she argued, would help protect against the industry's tendencies toward monopoly. Competition for contracts is often restrained by its intermittent and unsustained nature: It occurs only at the outset of long-term contracts, which are often renegotiated rather than recompeted upon expiration. Regulation would discipline competition and make it more predictable and accountable. It would also depoliticize investment and pricing decisions. Furthermore, regulation would facilitate restructuring in the industry: Levelling the playing field by imposing comparable standards on all utilities would make it easier to consolidate small and mid-size systems to achieve economies of scale. "In due course," she concluded, "regulation may be the best thing that could happen to privatization."[41]

Of course, much depends on the form that regulation takes. After participating for more than a decade in the regulation of oil, gas, and electricity utilities in Canada and examining various regulatory structures to determine what works and what creates problems, Mr. Adams recommended to the Walkerton Inquiry a set of principles to guide the economic regulation of water and sewage utilities.[42] Regulation, he advised, must be stable, reliable, and transparent. It must observe due process: Affected parties must have the right to participate fully, to access all relevant information, to cross examine other parties, and if necessary, to hold regulators accountable through judicial review. In order to ensure a balanced presentation of various points of view, public-interest intervenors should have the opportunity to recover their costs. These and other regulatory costs are properly included in the cost of service and should be borne by consumers. For regulation to be effective, it must also be manageable. Simplicity is important, especially for smaller utilities. Understanding that the regulatory process can be costly and difficult, many U.S. commissions have simplified filing and reporting procedures for smaller systems.[43] A formulaic approach may be appropriate for small Canadian systems, with the smallest governed by a complaints-based process in which consumers could initiate regulatory intervention.

However finely tuned, regulation will remain an imperfect substitute for competition. As the authors of the NRRI study explained, "Competitive markets are considered the best 'regulator' of economic behavior. Competition for market share and profits drives firms toward efficiency. Customers, including municipalities, are protected by their freedom to choose."[44] England and Wales are showing the world the many forms that competition can take in a fully privatized industry. In the absence of full privatization, competition generally takes the form of contracting out. Experience south of the border demonstrates the ample scope for competition for contracts. Although three large firms control almost 70 percent of the market, 13 other firms also compete in the national and regional markets, and hundreds of local firms run small systems.[45] Canada's limited experience suggests that rigorous competition for contracts is possible here, too. In 1996, Thunder Bay received 30 responses to its request for expressions of interest in making $70-million worth of upgrades to its water and wastewater plants. Two years later, 22 firms responded to Halifax-Dartmouth's request for expressions of interest in designing, building, and financing the construction of wastewater treatment facilities.[46] Even Goderich, Ontario – population 7,500 – attracted eight submissions in response to a request for qualifications to operate its water and wastewater facilities.

Despite the plethora of competitors, a number of factors impede competition. In Ontario, water and sewage contractors identify the Ontario Clean Water Agency (OCWA) as the biggest single barrier to a competitive environment. They complain that the public agency's advantages – including a municipal loan portfolio acquired at a discount, favourable tax treatments, freedom from generating profits or paying dividends, a provincial financial guarantee, a market share gained by government fiat rather than merit, and other advantages arising from its links to the provincial government – tilt the playing field in its favour and distort competition.[47] Some private firms fear that EPCOR, owned by the City of Edmonton, may likewise undercut competition. Although EPCOR operates more like a private company, the perception that it is government-backed may give it subtle competitive advantages. For example, the District of Port Hardy, on Vancouver Island, felt that EPCOR's status made competition unnecessary and negotiated a sole source contract with it.

The practice of sole sourcing is, of course, not limited to government-owned firms, and it is another barrier to a competitive market in Canada. A number of municipalities have sole-sourced their water or wastewater contracts to private firms. Some have appreciated the convenience, ease, and speed of sole-sourcing. Some have felt comfortable doing business with a familiar proponent or have wished to aid a favoured son.

(Hamilton, for example, wanted to give a local firm a start in the business.) Some have believed that one firm had a clear advantage over others. (In Tofield and Ryley, Alberta, the gas utility could lay a water pipeline in its existing right-of-way.) As the costs and risks of sole-sourcing are revealed, as the efficiency gains from competition are more widely recognized, as barriers to competition fall, and as municipalities acquire tools to help them navigate the competitive market, sturdier competition will doubtless emerge. Only then will municipalities realize the full benefits of privatization.

CONCLUSION
HOW FAR SHOULD WE GO?

"Privatization" has come to mean many different things to different people. As illustrated in the preceding chapters, the term encompasses a continuum of private sector involvement. It may signify the outright sale of assets, the granting of long-term concessions, the signing of short-term contracts, or even simply the opening of bidding on contracts that ultimately remain with government employees. The privatization of the water and wastewater systems in England and Wales stands at one extreme. At the other extreme stands the competitive process under which public-sector workers secured a five-year agreement to operate and maintain the water and wastewater plants in Charlotte, North Carolina.[1]

Each form of water and wastewater utility privatization has its own advantages. Because the various forms that privatization can take are still relatively new to most of the world, no single model has yet proven superior to others. What works best depends very much on the particular circumstances of the privatizing community.

In general, full privatization attracts the greatest investment of private capital. Assigning to the private sector responsibility for capital investments tends in turn to create the clearest accountability. Asset sales also enhance accountability by fully separating the owner of a system from its regulator, reducing the conflicts of interest that inhibit the regulation of public facilities. On the other hand, privatizing via competition for contracts may bring the greatest savings, with opportunities for further savings when the contracts expire and are again put up for competition. Competition has the additional advantage of generating savings internally: The savings are determined by the companies themselves rather than imposed by a regulator. Of course, not all contracts are created equal. Long-term contracts and concessions provide more opportunities for capital investment than their short-term counterparts and generally assign more responsibility to private operators. On the other hand, short-term contracts may be more readily embraced by cautious municipal councillors or voters. No arrangement holds a monopoly on attracting expertise or encouraging efficiency; even competitions for short-term contracts often draw experienced water and wastewater companies that improve performance and cut costs.

One increasingly common form of "privatization" does not make the grade. Under managed competition, a government agency or its workers compete with private bidders for a contract.[2] Managed competition clearly encourages efficiency in the public sector. However, even if public-sector workers are willing to reduce costs to a competitive level, managed competition does not bring with it several other important benefits associated with privatization, namely capital investment, technical expertise, or the elimination of the conflict of interest plaguing state-owned and - regulated facilities. It also presents unique challenges regarding accountability and the transfer of risk: Without shareholders to absorb losses, who but taxpayers or ratepayers will pay if government workers low-ball a bid or perform poorly? Who guarantees agreements with employees? How can they be enforced? Under managed competition, it is virtually impossible to hold the bidders themselves accountable.[3] For all of these reasons, while managed competition is better than no competition, it remains inferior to privatization.

Municipalities will have to examine their own needs, assess their privatization options, compare the gains from privatization to the status quo, and choose the model that will work best for them. In order to enable municipalities to weigh the real merits of different privatization options, the federal and provincial governments should remove artificial barriers to any particular option. The playing field on which different privatization options compete must be as level as possible. For example, tax treatments, subsidies, grant repayment provisions, or regulatory requirements should not tip the balance away from full privatization and towards contracting out.[4]

Whatever form of privatization a municipality choses, it is critical that it be subject to a good contract that unambiguously spells out the responsibilities of each party. Unclear language in contracts frequently creates disputes over which party should absorb particular costs. Arbitration or litigation may be time-consuming and costly and may damage the relationship between a municipality and its service provider.[5] It is far cheaper for a municipality to invest in a team of recognized experts – consultants and lawyers who can identify and specify its needs and draft a clear contract – than to later pay the consequences of a sloppy approach to privatization.

Even more important is the need for tough regulation. Privatization cannot mean deregulation. Water and wastewater utilities have enormous impacts on public health and the environment. The strict enforcement of effective laws is required to ensure that those impacts are positive. While the reduced conflicts of interest inherent in privatization facilitate law enforcement, they do not by any means guarantee it. Governments have to be deeply committed to enforcing the law, and citizens have to remain vigilant and step in when governments fail to act. Citizens must also have the opportunity to be involved in the economic regulation of water and wastewater utilities. Because such utilities generally hold monopolies over essential public services, both their services and prices have to be tightly controlled. Effective economic regulation remains one of the great challenges of privatization.

Back in 1994, when England's water regulator was still finding its feet, and prices, share values, and dividends were soaring, *Financial Times* columnist Joe Rogaly issued a passionate plea for strict regulation of the water and wastewater companies: "State monopolies are thoroughly discredited. We know their faults. But when you turn to private monopolists to do a job, you regulate them with chains, and where necessary beat them with rods, lash them, kick them and confine them. Do all of that, or they will take you for everything you've got."[6] Mr. Rogaly was right. Private water and wastewater monopolies cannot and will not regulate themselves. That is government's job. Indeed, it is one job that only government can do.

ENDNOTES

INTRODUCTION

1   William L. Megginson, Robert C. Nash, and Matthias van Randenborgh, "The Financial and Operating Performance of Newly Privatized Firms: An International Empirical Analysis," in *The Privatization Process: A Worldwide Perspective*, ed. Terry L. Anderson and Peter J. Hill (Lanham, Maryland: Rowman & Littlefield, 1996), p. 127.

2   Mary M. Shirley, *Trends in Privatization* (Washington: Center for International Private Enterprise, 1998), pp. 1- 3; and Robert W. Poole Jr., "Privatization for Economic Development," in *The Privatization Process, op. cit.,* p. 16.

3   Megginson et al., *op. cit.,* p. 116.

4   Poole, "Privatization for Economic Development," *op. cit.,* p. 1.

5   Robert W. Poole Jr., Summary of speech to Canadian Council for Public-Private Partnerships, Toronto, May 18, 1999 [online] [consulted October 3, 2001] <http://www.pppcouncil.ca/poole.pdf>

6   John Nesbitt, cited by Robert W. Poole Jr., "The Global Privatization Revolution," Policy Fax Analysis (Glastonbury, Connecticut: The Yankee Institute, 1995). Adapted from a speech delivered to a conference on privatization in Hartford, Connecticut, sponsored by the Yankee Institute and Trinity College, September 1994.

7   National Round Table on the Environment and the Economy, *NRTEE Sustainable Cities Initiative: Final Report and Recommendations*, 1999 [online] [consulted October 3, 2001] <http://www.nrtee-trnee.ca/eng/programs/ArchivedPrograms/Sustainable_Cities/report_complete.htm>

8   *Public Works Financing: 2000 International Major Projects Survey*, Vol. 144, October 2000, pp. 1, 8.

9   Poole, "Privatization for Economic Development," *op. cit.*, p. 16.

10   John Barham, Financial Times, "Global water industry requires 'radical change,'" *Financial Post*, October 8, 1997.

11   Carl Myers, "Privatisation: value for money?" *World Water and Environmental Engineering*, March 1997, Vol. 20, No. 3, p. 10.

12   *Public Works Financing: 1999 International Major Projects Survey*, pp. 8-10, and *Public Works Financing: 2000 International Major Projects Survey*, pp. 10-12, summarize recent private water and wastewater projects in 67 countries. *Public Works Financing's* monthly reports provide greater detail on many of the transactions. Other privatizing countries are discussed in Alex Orwin, *The Privatization of Water and Wastewater Utilities: An International Survey* (Toronto: Environment Probe, 1999); *World Water and Environmental Engineering* (monthly issues); Fitzroy Nation and Andrew Nickson, *Liquid Assets: Is water privatization the answer to access?*, Panos Media Briefing No. 29 (London: The Panos Institute, July/August 1998), pp. 7-8; and Frannie Leautier, "Public Private Partnerships in Infrastructure: Lessons from Developing Countries," presentation to *Narrowing the Gap*, conference sponsored by Canadian Council for Public-Private Partnerships, November 27, 2000.

13   Lyonnaise des Eaux, "Lyonnaise des Eaux Signs the Water Management Contract of Johannesburg," Press release, February 14, 2001; and Vivendi Environment, *2000 Environmental Report* [online] [consulted October 3, 2001] <http://www.vivendiuniversal.com/txt/vu/en/fichiers/envt2000_ang.pdf>

14   "Havana water to Agbar," *Public Works Financing*, Vol. 126, February 1999, pp. 8-9; and "Cuba taps Spain for water concession," *Public Works Financing*, Vol. 132, September 1999, p. 5.

15   "ADB loan for Chengdu water" and "Lyonnaise/New World in Chinese water," *Public Works Financing*, Vol. 128, April 1999, p. 25; "Road King/Anglian eye China water," *Public Works Financing*, Vol. 132, September 1999, p. 15; and "Beijing water RFP," *Public Works Financing*, Vol. 132, September 1999, p. 15.

16   "Vietnamese water deal agreed" and "Binh An BOT water plant start-up," *Public Works Financing*, Vol. 131, July/August 1999, pp. 22, 23.

17   United Nations Economic and Social Council, *Progress Made in Providing Safe Water Supply and Sanitation for All During the 1990s*, Doc. No. E/CN.17/2000/13, March 14, 2000, pp. 7, 13. In contrast, the Asian Development Bank estimates that 3 billion people lack adequate sanitation. Asian Development Bank, "Special Theme: Water in the 21st Century," in *Annual Report 1999*, p. 13.

18   Asian Development Bank, *op cit*. In contrast, the World Health

Organization estimates that 9,300 people die every day from water-related diseases. World Health Organization, "Water – Immediate Solutions for Persistent Problems," Press release, March 21, 2001.

19   William J. Cosgrove and Frank R. Rijsberman, for the World Water Council, *World Water Vision: Making Water Everybody's Business* (London: Earthscan, 2000), p. xxv. In contrast, Unicef estimates that US$25 billion a year will be required for 10 years – a sum that it puts at three times current expenditures. Unicef, *Sanitation for All: Promoting Dignity and Human Rights*, January 2000, p. 7.

20   David Owen, Samer Iskandar, and Andrew Taylor, "Making a big splash," *Financial Times*, August 24, 1999.

21   Emanuel Idelovitch and Klas Ringskog, *Private Sector Participation in Water Supply and Sanitation in Latin America* (Washington: The World Bank, May 1995), p. 35; and Metropolitan Waterworks and Sewerage System, "The ten most-asked questions about the privatization of the MWSS."

22   Nation and Nickson, *op. cit.*, p. 6.

23   Frederic Niel, Reuters News Agency, "U.N. summit says clean water should be a commodity," *Toronto Star*, April 4, 1998.

24   Sue Bailey, Canadian Press, "Dirty water killing millions every year, study warns," *Globe and Mail*, August 6, 1999.

25   World Commission for Water in the 21st Century, *A Water Secure World: Vision for Water, Life, and the Environment*, World Water Vision Commission Report, March 2000, p. 34.

26   Peter Allison, "Getting the price right: thoughts on Paris," *World Water and Environmental Engineering*, Vol. 21, No. 5, May 1998, p. 3.

27   This phrase, coined by E.S. Savas, was popularized by David Osborne and Ted Gaebler in *Reinventing Government* (Reading, Massachusetts: Addison-Wesley, 1992).

CHAPTER ONE

1   Janice A. Beecher, G. Richard Dreese, and John D. Stanford, *Regulatory Implications of Water and Wastewater Utility Privatization* (Columbus, Ohio: The National Regulatory Research Institute, July 1995), p. 21.

2   David T. Beito, "From Privies to Boulevards: The Private Supply of Infrastructure in the United States during the Nineteenth Century," in *Development by Consent: The Voluntary Supply of Public Goods and Services*, ed. Jerry Jenkins and David E. Sisk (San Francisco: Institute for Contemporary Studies Press, 1993), pp. 25-6, 46-7.

3   David Haarmeyer, *Privatizing Infrastructure: Options for Municipal Water-Supply Systems*, Policy Study No. 151 (Los Angeles: Reason Foundation, October 1992), p. 4.

4   United States Environmental Protection Agency, *1995 Community Water System Survey: Overview, Volume I*, EPA #815R97001A, pp. 7-8.

5   United States Environmental Protection Agency, *Small Water System Characteristics*, prepared by the Cadmus Group using data from *1995 Community Water System Survey*, draft dated January 7, 1999.

6   Haig Farmer, United States Environmental Protection Agency, E-mail to Elizabeth Brubaker, January 21, 2000.

7   Beecher, Dreese, and Stanford, *op. cit.*, p. 23.

8   Michael Ancell, "Privatization of Public Facilities: Panacea or Pipe Dream?" *The National Utility Contractor*, March 1993. Reprinted by the United States Environmental Protection Agency, May 1993, p. 3.

9   William G. Reinhardt, "U.S. Water/Wastewater Contract Services Show 16% Growth to $1.7 Billion in 2000," *Public Works Financing*, Vol. 149, March 2001, p. 2.

10   United States Environmental Protection Agency, *Liquid Assets: A Summertime Perspective on the Importance of Clean Water to the Nation's Economy*, EPA #800-R-96-002, May 1996, p. v.

11   United States Environmental Protection Agency, *Providing Safe Drinking Water in America: 1996 National Public Water System Annual Compliance Report*, EPA #305/R-98-001, September 1998.

12   United States Environmental Protection Agency, *Liquid Assets 2000: America's Water Resources at a Turning Point*, EPA #840-B-00-001, May 2000.

13   Nicolas Spulber and Asghar Sabbaghi, *Economics of Water Resources: From Regulation to Privatization*, Second Edition (Boston: Kluwer Academic Publishers, 1998), pp. 200-02.

14   Holly June Stiefel, *Privatization of Municipal Wastewater Treatment Plants: Effect on Responsibility and Compliance*, Masters Thesis, Penn State Environmental Pollution Control Program, May 1992, pp. 6-8.

15   United States Environmental Protection Agency, *Liquid Assets 2000*, *op cit.*

16   Mark Dorfman, *Testing the Waters 2001: A Guide to Water Quality at Vacation Beaches* (New York: Natural Resources Defense Council, August 2001).

17   Richard Caplan, *Polluters' Playground: How the Government Permits Pollution* (Washington: U.S. Public Interest Research Group Education Fund, May 2001), p. 5.

18   United States Environmental Protection Agency, Office of Wastewater Management, "Proposed Rule to Protect Communities

from Overflowing Sewers," EPA #833-01-F-001, January 2001.

19   Hudson Institute, *The NAWC Privatization Study: A Survey of the Use of Public-Private Partnerships in the Drinking Water Utility Sector* (Washington: National Association of Water Companies, June 1999), pp. 26, 31, 50.

20   United States Environmental Protection Agency, *Response to Congress on Privatization of Wastewater Facilities*, EPA #832-R-97-001a, July 1997, p. 7.

21   United States Environmental Protection Agency, *1999 Drinking Water Infrastructure Needs Survey: Second Report to Congress*, February 2001, p. 11; and "U.S. wastewater infrastructure needs $700Bn investment through 2015," *Edie Weekly Summaries*, April 16, 1999.

22   Water Infrastructure Network, *Clean and Safe Drinking Water for the 21st Century*, April 2000, p. ES-1.

23   Skip Stitt, in *Water Delivery Systems: An International Comparison*, Unedited transcript of panel discussion hosted by the Canadian Council for Public-Private Partnerships, Toronto, November 1998, p. 13.

24   Hudson Institute, *op. cit.*, executive summary and pp. 28-9.

25   Beecher, Dreese, and Stanford, *op. cit.*, pp. 67, 69, 72.

26   Holly June Stiefel, *Municipal Wastewater Treatment: Privatization and Compliance*, Policy Study No. 175 (Los Angeles: Reason Foundation, February 1994), pp. 6-9.

27   Jerry Ellig, *The $7.7 Billion Mistake: Federal Barriers to State and Local Privatization*, Joint Economic Committee Staff Report, February 1996, pp. 2-3.

28   "Govs, Mayors Shun Federal Water Grants," *Public Works Financing*, Vol. 148, February 2001, p. 8.

29   John Stokes, Meeting with Elizabeth Brubaker and Mark Hudson, February 9, 2001.

30   "The Right Price for Water," *Public Works Financing*, Vol. 127, March 1999, p. 17.

31   John-Thor Dahlburg, "Water Companies: Tap Water Around the World Developing a French Flavor," *Los Angeles Times*, April 30, 2000.

32   Stitt, *op. cit.*, p. 13.

33   Ancell, *op. cit.*, p. 1; and United States Environmental Protection Agency, *Response to Congress, op. cit.*, pp. 4, 8, 16, 21.

34   Spulber and Sabbaghi, *op. cit.*, p. 290.

35   Ancell, *op. cit.*, pp. 1-2; and United States Environmental Protection Agency, *Response to Congress, op. cit.*, pp. 1, 4, 23.

36   United States General Accounting Office, *Privatization: Lessons*

Learned by State and Local Governments, Report to the Chairman, House Republican Task Force on Privatization, GAO/GGD-97-48, March 1997, p. 2.

37   Nolan Bederman, "Franklin, Ohio and Wheelabrator E.O.S.: Public-Private Partnership for Wastewater Management," in *Case Studies in Public-Private Partnerships*, ed. Michael J. Trebilcock (Toronto: Canadian Council for Public-Private Partnerships, August 1996), pp. 41-2; and United States Environmental Protection Agency, *Response to Congress, op. cit.*, p. 4.

38   United States Environmental Protection Agency, *Response to Congress, op. cit.*, p. 2.

39   United States General Accounting Office, *op. cit.*, pp. 6-7, 26-30.

40   Stuart Butler, "Privatization for Public Purposes," in *Privatization and Its Alternatives*, ed. William T. Gormley, Jr. (Madison: The University of Wisconsin Press, 1991), p. 24.

41   National Association of Water Companies, *1998 Financial and Operating Data for Investor-Owned Water Utilities* (Washington: NAWC Economic Research Program, 1999).

42   Hudson Institute, *op. cit.*, pp. 35, 39.

43   United Water, *1999 Annual Report*, p. 13.

44   "United Water Extends Hoboken, N.J.," *Public Works Financing*, Vol. 153, July-August 2001, p. 9.

45   Stitt, *op. cit.*, p. 22.

46   The relative efficiency of publicly and privately owned or operated water utilities has been subject to much debate, with no resolution that satisfies all sides. Despite the well-developed and persuasive theories arguing that incentive structures and other factors inherent in competitive or well-regulated private water regimes encourage efficiency, studies comparing public and private utilities have yielded mixed results.

Beecher, Dreese, and Stanford (*op. cit.*, pp. 34-7) reviewed 13 studies, conducted between 1971 and 1994, comparing public and private water utilities. Four of the studies found public firms to be more efficient, four found private firms to be more efficient, and five found no significant difference or obtained ambiguous results.

Lambert, Dichev, and Raffiee reviewed five studies, most of which found little difference. Their own calculations, however, suggested that public water utilities were more efficient. They identified the overuse of capital as the greatest source of inefficiency. (David K. Lambert, Dimo Dichev, and Kambiz Raffiee, "Ownership and Sources of Inefficiency in the Provision of Water Services," *Water Resources Research*, Vol. 29, No. 6, June 1993, pp. 1573-8.)

Unsurprisingly, different cost models have yielded different results. Teeples and Glyer found that their dual cost function model reduced efficiency differences to insignificance. (Ronald Teeples and David Glyer, "Cost of Water Delivery Systems: Specification and Ownership Effects," *Review of Economics and Statistics*, 1987, pp. 399-408.)

What is beyond dispute is that experience with the privatization process in the last decade demonstrates that private owners and operators are often more efficient than their public predecessors: In many cases, following privatization, utilities provide comparable service at far lower cost or dramatically improve service without increasing cost. That some of these efficiencies may be attributable to increased competition or enhanced regulation rather than to ownership *per se* does not diminish the role that the privatization process plays in bringing them about. This book probes some of the contributing factors in an effort to understand how and why efficiencies may be achieved.

47 Harry Kitchen, *Efficient Delivery of Local Government Service*, Government and Competitiveness Project, School of Policy Studies, Queen's University Discussion Paper No. 93-15, pp. 12-13.

48 David Haarmeyer, *Privatizing Infrastructure: Options for Municipal Water-Supply Systems*, Policy Study No. 151 (Los Angeles: Reason Foundation, October 1992), pp. 26-7.

49 Kathy Neal, Patrick J. Maloney, Jonas A. Marson, and Tamer E. Francis, *Restructuring America's Water Industry: Comparing Investor-Owned and Government-Owned Water Systems*, Policy Study No. 200 (Los Angeles: Reason Foundation, January 1996), pp. 3, 10.

50 Adrian Moore, *Clearing Muddy Waters: Private Water Utilities Lower Costs and Improve Services* (Los Angeles: Reason Public Policy Institute, 1997), p. 1.

51 Reason Public Policy Institute, "Wastewater Treatment," *Privatization Database* [online] [consulted December 16, 1999] <http://www.privatization.org>

52 "Tampa Bay Water DBO Award," *Public Works Financing*, Vol. 139, April 2000, p. 5; "Newport, R.I., Votes for Earth Tech,"*Public Works Financing*, Vol. 146, December 2000, p. 7; "Water/Wastewater Privatization Hits a Growth Spurt," *Public Works Financing*, Vol. 127, March 1999, p. 17; "IRS Rules in Favor of Pension Portability," *Public Works Financing*, Vol. 129, May 1999, p. 2; "Water Privatization Scorecard," *Public Works Financing*, Vol. 126, February 1999, p. 8; "Market Awaits Enron's Answer," *Public Works Financing*, Vol. 127, March 1999, p. 3; and "Tampa Bay Water Plant DBO," *Public Works Financing*, Vol. 129, May 1999, p. 8.

53 Ancell, *op. cit.*, pp. 2-3.
54 Thompson Gow and Associates, *Canada's Untapped Resource: Public-Private Partnerships in Water Supply and Wastewater Treatment,* Prepared for Technology Transfer Office, Environmental Technology Advancement Directorate, Environment Canada, Technology Transfer Series 2E, September 1995, pp. 48-50.
55 Discussion between water industry representative and Elizabeth Brubaker, April 2001.
56 Hudson Institute, *op. cit.*, p. 37.
57 United States Environmental Protection Agency, *Response to Congress, op. cit.*, pp. 33-4.
58 "NJ BPU to Rule on Concession Fee," *Public Works Financing,* Vol. 137, February 2000, p. 5.
59 "Scranton, Pa., O&M Sole-Sourced," *Public Works Financing,* Vol. 127, March 1999, p. 9; "Rahway, N.J, Water Concession," *Public Works Financing,* Vol. 127, March 1999, p. 10; and "Rahway, N.J. Water Fees Back- Loaded," *Public Works Financing,* Vol. 132, September 1999, p. 14.
60 Diane Kittower, "Serving the Public with Private Partners," Guide to Privatization, *Governing* (published by Congressional Quarterly Inc.), May 1997, p. 74.
61 Daniel J. Kucera, "A Return to Franklin: How is Privatization Working?" *Public Works Online,* April 16, 1999; and United States Environmental Protection Agency, *Response to Congress, op. cit.*, p. 31.
62 Adrian Moore, Geoffrey Segal, John McCormally, *Infrastructure Outsourcing: Leveraging Concrete, Steel, and Asphalt with Public-Private Partnerships,* Policy Study No. 272 (Los Angeles: Reason Public Policy Institute, September 2000). p. 22.
63 Moore et al., *op. cit.*, pp. 22-3.
64 Hudson Institute, *op. cit.*, pp. 27, 36, 42.
65 Neal et al., *op. cit.*, pp. 4-5, 9-10, 14-15.
66 Hudson Institute, *op. cit.*, pp. 35, 39.
67 Kucera, *op. cit.*
68 Milwaukee Metropolitan Sewerage District, "Terms of the 10-year Contract," *Competitive Contracting News,* March 1998, p. 4; Milwaukee Metropolitan Sewerage District, *Competitive Contract Annual Report,* March 1999, p. 5; Milwaukee Metropolitan Sewerage District, *Competitive Contract Annual Report,* March 2000; and Milwaukee Metropolitan Sewerage District, *Competitive Contract 2001 Annual Report,* 2001 [online] [consulted October 24, 2001] <http://www.mmsd.com/cmpcontract00/comp2000.asp>
69 "FBI's PSG Probe Ripens," *Public Works Financing,* Vol. 148, February

2001, p. 3; "Fines for New Orleans Water Bribe," *Public Works Financing*, Vol. 151, May 2001, p. 5; "Old Orleans Graft a ConOps Waterloo," and "FBI Net Spreads in Bridgeport-PSG Probe," *Public Works Financing*, Vol. 152, June 2001, pp. 4-5.

70 Carissa Mire, "Trial date scheduled," *Daily Iberian*, November 4, 2001; and Richard Burgess, "Ex-Jeanerette mayor's trial set for Dec. 3," *Daily Advertiser*, November 6, 2001.

71 Janice Beecher, "Privatization, Monopoly, and Structured Competition in the Water Industry: Is There a Role for Regulation?" in Universities Council on Water Resources, *Issue No. 117: Private Sector Participation in Urban Water Supply*, October 2000, p. 14.

72 City of Indianapolis, "Frequently Asked Questions: Public Ownership of the Indianapolis Water Company," undated; Doug Sword, "City says water utility sale averts big rate increase," *Indianapolis Star*, December 19, 2001; and NiSource Inc., "NiSource Completes Sale of Indianapolis Water Company," Press release, April 30, 2002.

73 USFilter, "City of Indianapolis Selects USFilter to Manage Waterworks System," Press release, March 19, 2002.

74 "Renewals, Lost Contracts for 13 Firms in 1998," *Public Works Financing*, Vol. 127, March 1999, p. 24.

75 "Renewals, Lost Contracts in Calendar 2000," *Public Works Financing*, Vol. 149, March 2001, p. 21.

CHAPTER TWO

1 Charmagne Helton, "Focus on Sewer Privatization: Atlanta Looks to Indianapolis for Guidance," *Atlanta Journal / Atlanta Constitution*, April 1, 1997.

2 City of Atlanta, *Atlanta Water Commissioner's Updates*, November 21, 1997.

3 "Atlanta: All eyes watch country's biggest privatization effort," *American City and County*, May 1998.

4 "Campbell bridges troubled water issue," *Atlanta Business Chronicle*, September 4, 1998.

5 Charmagne Helton, "Focus on Sewer Privatization," *op. cit.*

6 Richard Lambert, "Investors: Come Splash Out in Atlanta's Water," *Financial Times*, March 6, 1998.

7 Stephanie Ramage, "Campbell keeps battling," *Atlanta Business Chronicle*, September 4, 1998.

8 According to Atlanta's *City Beat Online*, the city anticipated an increase

of 81 percent. City of Atlanta, "Privatization Milestones," *City Beat Online*, Vol. 1, No. 14, August 22, 1998 [online] [consulted April 4, 2001] <http://www.ci.atlanta.ga.us/Citybeat/aug2298/page3.htm>

9   Bill Campbell, "Why I believe privatization is the right thing to do," *Atlanta Constitution*, February 1998. Reprinted in City of Atlanta, *City Beat Online*, Vol. 1, No. 14, August 22, 1998 [online] [consulted February 28, 2001] <http://www.ci.atlanta.ga.us/Citybeat/aug2298/page2.htm>

10   James Waddell, *Public Water Suppliers Look to Privatization*, White Paper included in Montgomery Research's Utilities Project, undated [online] [consulted April 3, 2001] <http://www.utilitiesproject.com>

11   Bill Campbell, "Last Call for Privatization," Press conference, August 19, 1998. Posted on City of Atlanta, *City Beat Online*, Vol. 1, No. 15, August 29, 1998 [online] [consulted April 4, 2001] <http://www.ci.atlanta.ga.us/Citybeat/aug2998/aug2998.htm>

12   "Atlanta: All eyes watch country's biggest privatization effort," *op. cit.*; and United Water, "Atlanta signs United Water," undated [online] [consulted April 3, 2001] <http://waterindustry.org>

13   "Atlanta: All eyes watch country's biggest privatization effort," *op. cit.*

14   Andrew Young, "Statement on Privatization and the Metro Group," *City Beat Online*, Vol. 1, No. 14, August 22, 1998 [online] [consulted February 28, 2001] <http://www.ci.atlanta.ga.us/Citybeat/aug2298/aug2298.htm>; and City of Atlanta, "Privatization a Done Deal," *City Beat Online*, Vol. 1, No. 23, October 24, 1998 [online] [consulted April 4, 2001] <http://www.ci.atlanta.ga.us/Citybeat/Oct2498/Oct2498.htm>

15   Ramage, *op. cit.*

16   Jess Scheer, "Water privatization: Oversight firm tied to 'crony,'" *Creative Loafing*, August 22, 1998; and Jess Scheer, "Anonymous letter says water contract bid 'rigged,'" *Creative Loafing*, August 15, 1998.

17   Ramage, *op. cit.*

18   United Water, "Atlanta signs United Water," *op. cit.*

19   United Water, *1999 Annual Report*, pp. i, 15.

20   Bill Campbell, "Why I believe privatization is the right thing to do," *op. cit.*; and Carlos Campos, Julia Hairston, and David Pendered, "United Water 'a safe selection,'" *Atlanta Journal-Constitution*, August 8, 1998.

21   "Privatization a Done Deal," *op. cit.*

22   Adrian Moore, *Privatization 1999: The 13th Annual Report on Privatization* (Los Angeles: Reason Foundation, May 1999), p. 36.

23   Jeff Dickerson, "Ambitious Commitment: Empowerment Zone has new best friend," *Atlanta Journal- Constitution*, December 1, 1998; and City of Atlanta, "Water Privatization Floats Hope in Empowerment Zone," *City*

Beat Online, Vol. 1, No. 20, October 3, 1998 [online] [consulted April 4, 2001] <http://www.ci.atlanta.ga.us/Citybeat/Oct0398/Oct0398.htm>
24 U.S. Conference of Mayors, "2001 Public/Private Partnership Awards" [online] [consulted October 29, 2001] <http://usmayors.org/USCM/best_practices/buscouncil/atlanta.asp>
25 "Drinking Success: Atlanta Project in Capsule" in Water Industry Council Information Archives [online] [consulted April 3, 2001] <http://waterindustry.org/frame-8.htm>
26 United Water, *Municipal Info: Atlanta, GA*, undated [online] [consulted March 7, 2001] <http://www.unitedwater.com/atlanta.htm>
27 United Water, *1999 Annual Report*, p. 13; and United States Conference of Mayors, "Best Practice Recognition: The City of Atlanta and United Water Services Atlanta Privatization of Water Services," Best Practices Database [online] [consulted April 4, 2001] <http://www.usmayors.org/uscm/best_practices/partnership/atlanta_reg.html>
28 Moore, *op. cit.*, p. 36.
29 City of Atlanta, "Administration Selects United Water Services to Run City's Water System," *City Beat Online*, Vol. 1, No. 16, September 5, 1998 [online] [consulted April 4, 2001] <http://www.ci.atlanta.ga.us/Citybeat/sep0598/sep0598.htm>; "Privatization a Done Deal," *op. cit.*; and United Water, *Municipal Info: Atlanta, GA*, undated [online] [consulted March 7, 2001] <http://www.unitedwater.com/atlanta.htm>
30 Waddell, *op. cit.*
31 "Privatization a Done Deal," *op. cit.*; and American Federation of State, County, and Municipal Employees, "No layoffs for 20 years," *AFSCME Leader*, June 1999 [online] [consulted March 7, 2001] <http://www.afscme.org/publications/leader/99060110.htm>
32 According to one newspaper account, the water department had 731 employees two years before privatization. The number declined as workers retired or found new jobs. Ann Hardie, "City, firm up to their necks in complaints," *Atlanta Journal-Constitution*, September 6, 1999.
33 United Water, "Atlanta signs United Water," *op. cit.*; and United Water, *Municipal Info: Atlanta, GA, op. cit.*
34 "United Gets Good Marks on Atlanta Water Report Card," *Public Works Financing*, Vol. 141, June 2000, p. 4.
35 United Water, *Municipal Info: Atlanta, GA, op. cit.*; American Federation of Labor – Congress of Industrial Organizations, "Going with the flow," *Work in Progress*, April 19, 1999 [online] [consulted March 7, 2001] <http://www.aflcio.org/publ/wip1999/wip0419.htm>;

American Federation of State, County, and Municipal Employees, *op. cit.*; "Unions Sign Up for Atlanta Water," *Public Works Financing*, April 1999, Vol. 128, pp. 14- 15; Louis Monteilh, Telephone conversation with Krystyn Tully, April 11, 2001; Doug Reichlin, Telephone conversation with Elizabeth Brubaker, April 4, 2002; and Michelle Stewart Ware, Telephone conversation with Elizabeth Brubaker, April 8, 2002.

36 Carlos Campos, "Atlanta water flows to United: New Jersey firm optimistic despite employee lawsuit," *Atlanta Journal-Constitution*, January 1, 1999.

37 Gopal Tiwari, "Privatization of PEs good for public," *Kathmandu Post*, June 7, 2000.

38 Ann Hardie, "City, firm up to their necks in complaints," *op. cit.*

39 Ann Hardie, "Backlog damming tries at timeliness," *Atlanta Journal-Constitution*, August 30, 1999.

40 "United Gets Good Marks on Atlanta Water Report Card," *Public Works Financing*, Vol. 141, June 2000, pp. 3-4; and Robin Johnson and Adrian Moore, *Opening the Floodgates: Why Water Privatization Will Continue*, Policy Brief 17 (Los Angeles: Reason Public Policy Institute, August 2001).

41 United Water Services Atlanta, *Atlanta's 1998 Drinking Water Quality Report*; United Water Services Atlanta, *Atlanta Water System Water Quality Report for 1999*; and United Water Services Atlanta, *Atlanta Water System Water Quality Report for 2000*.

42 United Water, *1999 Annual Report*, p. 13.

43 Bill Campbell, "Building the New Atlanta Through Effective Partnerships," 2000 State of the City Millennium Business Breakfast [online] [consulted April 5, 2001] <http://www.ci.atlanta.ga. us/mayor/2000%20Speech/Page2.htm>

44 City of Atlanta, "Accomplishments of Mayor Bill Campbell," *City Beat Online*, Vol. 3, No. 49, June 9, 2001 [online] [consulted October 29, 2001] <http://www.ci.atlanta.ga.us/pdf/V3%20-%20No.%2049.pdf>

45 City of Atlanta, "Mayor Shirley Franklin Initiates Review of United Water Contract," Press release, June 11, 2002.

46 Rich Henning, Telephone conversation with Elizabeth Brubaker, July 29, 2002; and Doug Reichlin, Telephone conversation with Elizabeth Brubaker, August 8, 2002.

47 Alfred Charles, "Fulton, Atlanta Names OMI to run Northside waste water system," *Atlanta Journal- Constitution*, April 21, 2000.

48 "Azurix Wins Atlanta DBO for American," *Public Works Financing*, November 2001, Vol. 156, p. 4.

49 The following discussion is largely drawn from Craig Golding, *A Wastewater Treatment Privatization Case Study: Indianapolis, Indiana* (Toronto: Environment Probe, March 2000).

50 City of Indianapolis, "Goldsmith Announces Savings from Wastewater Privatization," Press release, August 5, 1999, p. 2.

51 Stephen Goldsmith, *The Twenty-First Century City: Resurrecting Urban America* (Washington, D.C.: Regnery Publishing, Inc., 1997), p. 35.

52 Goldsmith, *op. cit.*, p. 202.

53 Jeff Bowden, Glena Carr, and Judi Storrer, *New Directions in Municipal Services: Competitive Contracting and Alternative Service Delivery in North American Municipalities* (Toronto: ICURR Press, 1997), p. 6.

54 White River Environmental Partnership, *City of Indianapolis Contract Operations of the AWT Facilities and Collection System: Fifth Year Summary of Activities*, undated, p. 7.

55 *Amended and Restated Agreement for the Operation and Maintenance of the City of Indianapolis, Indiana, Advanced Wastewater Treatment Facilities*, contract between the City of Indianapolis and White River Environmental Partnership, amended December 1997, pp. 15, 30.

56 Other sources put the number of city employees at between 321 and 328.

57 White River Environmental Partnership, *Indianapolis Advanced Wastewater Facilities: One Year Summary*, submitted to Indianapolis City County Council March 20, 1995, p. 9.

58 White River Environmental Partnership, *City of Indianapolis Contract Operations of the AWT Facilities and Collection System: Fourth Year Summary of Activities*, undated, p. 1.

59 WREP calls its switch to chlorination/dechlorination temporary. It defends it on the grounds that the ozone system could not handle large flows, broke down repeatedly, and required costly repairs. It has proposed moving to a disinfection system that uses ultraviolet radiation and awaits city approval to do so. While many environmentalists oppose the use of chlorine, they see the switch as an acceptable temporary solution. As the Audubon Society's Richard van Frank explains, "The O3 system was nearly beyond repair and was down much of the time . . . Chlor/dechlor was the best immediate solution to the problem. UV would be better but, again, the city will not install it because of capital cost." (E-mail to Craig Golding, December 4, 1999.)

60 White River Environmental Partnership, *Indianapolis . . . : One Year Summary, op. cit.*, p. 4.

61 White River Environmental Partnership, *Indianapolis Advanced Wastewater Treatment Facilities: Second Year Summary*, undated, p. 4.

62 City of Indianapolis, "Goldsmith Announces Savings," *op. cit.*, p. 1 and Attachment D: WREP Savings Investments 1994-1998.

63 United Water, "United Water Contract Extended for Indianapolis Partnership," Press release, November 5, 1997.

64 *Amended and Restated Agreement*, pp. 18, 23, 24, 26, 27; and Daniel Mullins and C. Kurt Zorn, "Privatization in Indianapolis: A Closer Look at Savings and the Wastewater Treatment Facility," *In Roads*, Vol. 1, No. 3, Fall 1996, endnote 22, p. 20.

65 Charmagne Helton, "Focus on Sewer Privatization," *op. cit.*

66 White River Environmental Partnership, *Indianapolis . . . : One Year Summary, op. cit.*, p. 1.

67 City of Indianapolis, "Goldsmith Announces Savings," *op. cit.*, p. 3; and White River Environmental Partnership, *City of Indianapolis . . . : Fifth Year Summary of Activities, op. cit.*, p. 2.

68 White River Environmental Partnership, *City of Indianapolis . . . : Fifth Year Summary of Activities, op. cit.*, p. 2.

69 Charmagne Helton, "New Management Gets High Marks," *Atlanta Journal / Atlanta Constitution*, April 1, 1997.

70 David Osborne, "The Service Secret," *Washington Post Magazine*, December 8, 1996, p. 12.

71 Goldsmith, *op. cit.*, p. 208.

72 Kyle Niederpruem, "City, State Strike Deal Over '94 Fish Kill," *Indianapolis Star / News*, February 2, 1997.

73 Steven Cohen and William Eimicke, *Is Public Entrepreneurship Ethical? A Second Look at Theory and Practice*, School of International and Public Affairs, Columbia University, draft of May 18, 1998 [online] [consulted April 6, 2001] <http://www.columbia.edu/cu/sipa/COURS-ES/PUBMAN/pm4.html>

74 Glenn Pratt, E-mail to Craig Golding, August 3, 1999.

75 *Amended and Restated Agreement, op. cit.*, pp. 14, 19; and City of Indianapolis, "Goldsmith Announces Savings," *op. cit.*, p. 3.

76 City of Indianapolis, "Goldsmith Announces Savings," *op. cit.*, Attachment B: Effluent Performance Scorecard; and White River Environmental Partnership, *2000 Summary of Activities*, p. 3.

77 City of Indianapolis, *2000 Annual Budget: Department of Public Works Contract Compliance Division* [online] [consulted April 6, 2001] <http://www.indygov.org/controller/budget2000/dpw_budget.pdf >

78 City of Indianapolis, "Goldsmith Announces Savings," *op. cit.*, Attachment C: Awards.

79 Diana Penner, "City Sewage Plants Go Private to Cut Costs," *Indianapolis Star / News*, November 13, 1993.

80 Pratt, *op. cit.*

81 Van Frank, *op. cit.*

82 Brant Cowser, E-mail to Craig Golding, December 4,1999.

83 Reason Foundation, "Indianapolis Wastewater Contract Praised by Public Officials," *Privatization Watch*, No. 221, May 1995, p. 1.

84 City of Indianapolis, "Goldsmith Announces Savings," *op. cit.*, p. 1.
85 Dennis Neidigh, "Indianapolis Continues Role as Model for Public-Private Partnership," *Public Works*, March 1999, p. 55.

CHAPTER THREE

1   Office of Water Services, *Leakage of Water in England and Wales*, May 1996, pp. 4, 11, 13.
2   Donna Burnell, ed., *Waterfacts '95* (London: Water Services Association of England and Wales, 1995), p. 31.
3   David Kinnersley, *Coming Clean: The Politics of Water and the Environment* (London: Penguin Books, 1994), pp. 4, 49, 60-1.
4   The Secretary of State for the Environment et al., *Privatisation of the Water Authorities in England and Wales*, Government White Paper, February 1986, pp. 1, 2.
5   Kinnersley, *op. cit.*, p. 202; and Lord Crickhowell, *House of Lords Hansard*, April 17, 1989, col. 579.
6   Kinnersley, *op. cit.*, p. 53.
7   Office of Water Services, *Privatisation and History of the Water Industry*, Information Note No. 18, February 1993.
8   Shaoul compares the assets' sale price of £5.25 billion to their net book value of £8.87 billion. He also notes that the sale price was less than the debt write-off and the green dowry, concluding that, by any measure, the sale "constituted a loss to the Government and taxpayers." (Jean Shaoul, "A Critical Financial Analysis of the Performance of Privatised Industries: The Case of the Water Industry in England and Wales," *Critical Perspectives on Accounting* (1997) 8, p. 486.)
    The former Director General of Water Services points out that while taxpayers may have lost initially, they recouped their losses through the windfall tax levied in 1997. (Ian Byatt, "The Water Regulation Regime in England and Wales," in *Regulation of Network Utilities: The European Experience*, eds. Claude Henry, Michel Matheu, and Alain Jeunemaitre (Oxford: Oxford University Press, 2001), p. 81.)
9   Brian Cochrane, "Water fight," Letter to the Editor, *National Post*, March 9, 1999.
10 Canadian Union of Public Employees, *Who's Pushing Privatization? CUPE's 2000 Annual Report on Privatization* (Ottawa: CUPE, 2000), pp. vi, 16.
11 Canadian Union of Public Employees, "Canada's Water Under Threat," Water Watch campaign document, 1999.

12 Canadian Union of Public Employees, "Helpful Tips on Political Action," Water Watch campaign document, 1999.
13 Ontario Standing Committee on Resources Development, *Hansard*, April 14, 1997, 940.
14 Ontario Standing Committee on Resources Development, *Hansard*, April 14, 1997, 1040, 1050.
15 Ontario Standing Committee on Resources Development, *Hansard*, April 15, 1997, 1336.
16 Ontario Standing Committee on Resources Development, *Hansard*, April 15, 1997, 940, 1130.
17 Ontario Standing Committee on Resources Development, *Hansard*, April 16, 1997, 1520; April 14, 1997, 1200; and April 15, 1997, 1350.
18 Ontario Standing Committee on Resources Development, *Hansard*, April 16, 1997, 1430.
19 Ontario Standing Committee on Resources Development, *Hansard*, April 14, 1997, 1130; April 16, 1997, 1200; and April 30, 1997, 1630.
20 Maude Barlow, "Canada on Tap," *Canadian Perspectives*, Winter 1999.
21 Canadian Environmental Law Association, "Ontario's Push for Privatization," Backgrounder, May 29, 1998.
22 Ontario Public Service Employees Union, *Focus on Water Privatization: An Insiders' Report*, undated.
23 Ontario Standing Committee on Resources Development, *Hansard*, April 15, 1997, 1336; and April 16, 1997, 1640.
24 Ontario Standing Committee on Resources Development, *Hansard*, April 16, 1997, 1030.
25 Office of Water Services, *1998-99 Report on the Financial Performance and Expenditure of the Water Companies in England and Wales*, September 1999, p. 33.
26 Philip Fletcher, "The Future of the UK Water Regulatory Regime," speech to FT Global Water Conference, November 14, 2000.
27 Telephone conversation with Elizabeth Brubaker, March 18, 1997.
28 Sue Tabb, ed., *Waterfacts '97* (London: Water Services Association of England and Wales, 1997), p. 57.
29 Office of Water Services, *1998-99 Report on the Financial Performance*, p. 5.
30 Ian Byatt, "Life After the 1999 Periodic Review," Speech to Royal Aeronautical Society, January 26, 2000.
31 Caroline van den Berg, "Water Privatization and Regulation in England and Wales, *Public Policy for the Private Sector*, The World Bank, June 1997, p. 10.
32 David S. Saal and David Parker, "Productivity and Price Performance in the Privatized Water and Sewerage Companies of England and

Wales," *Journal of Regulatory Economics* 20:1, 2001, pp. 77-8.

33   Office of Water Services, *The Urban Waste Water Treatment Directive*, Information Note No. 24, revised June 1998; Tabb, *op. cit.*, p. 58; Water UK, *Waterfacts '98* (London: Water UK, 1998), p. 61; and Thompson Gow and Associates, *Canada's Untapped Resource: Public-Private Partnerships in Water Supply and Wastewater Treatment*, Prepared for Technology Transfer Office, Environmental Technology Advancement Directorate, Environment Canada, Technology Transfer Series 2E, September 1995, p. 30.

34   Ontario Standing Committee on Resources Development, *Hansard*, April 16, 1997, 1520; April 14, 1997, 1040; April 16, 1997, 1640, 1520; and April 14,1997, 1430.

35   Ontario Standing Committee on Resources Development, *Hansard*, May 5, 1997, 1640.

36   Environment Agency, "Water Companies: Things Can Still Get Better, Says Agency," Press release, September 1, 1999.

37   Environment Agency, *Achieving the Quality*, June 2000, p. 13.

38   Emer O'Connell, "Decade of clean-up brings best ever river and estuary quality results," Environment Agency press release, November 5, 2001.

39   Environment Agency, *State of the Environment: Freshwater Quality*, 1999.

40   Environment Agency, "Water Companies: Things Can Still Get Better," *op. cit.*; and Environment Agency, "Industrial Heartland Rivers Come Clean," Press release, September 21, 2000.

41   Environment Agency, *Spotlight on Business Environmental Performance: Report 2000*, September 2001, p. 11.

42   Environment Agency, *State of the Environment: Bathing Water Quality*, 1999; and Susanne Baker, "Wave goodbye to dirty beaches: Bathing water quality best since monitoring began," Environment Agency press release, November 7, 2001.

43   Environment Agency, "Water Companies: Things Can Still Get Better," *op. cit.*

44   Office of Water Services, *Leakage and Efficient Use of Water: 2000-2001 Report*, October 2001, p. 3.

45   Environment Agency, *Spotlight, op. cit.*, p. 19.

46   *Which?* [sic], "Pollution Can Be a Killer," undated [online] [consulted July 12, 2000] <http://www.which.net/publicinterest/freereport/03killer.html>

47   Ted Thairs, Fax to Elizabeth Brubaker, July 1, 1998.

48   "Environmental Infractions to Cost UK, Germany Dear," Environment News Service, December 22, 2000 [online] [consulted April 7, 2001]

<http://ens.lycos.com/ens/dec2000/2000L-12-22-11.html>
49　"Water quality good but could be better," Edie Weekly Summaries, November 27, 1998 [online] [consulted April 8, 2001] <http://www.edie.net/news/Archive/442.html>
50　"Utilities asked to act over polluted beaches," BBC online network, July 10, 1998.
51　Suzanne Baker, "Company Directors Must Show Zero Tolerance of Pollution: The Water Industry," Environment Agency press release, September 27, 2001.
52　"Water firms slammed over pollution," BBC online network, May 29, 1998.
53　Environment Agency, "Water Companies: Things Can Still Get Better," *op. cit.*; and Environment Agency, *Spotlight, op. cit.*, pp. 11, 13.
54　"Hall of Shame flushes out water polluters," Edie Weekly Summaries, March 26, 1999 [online] [consulted April 9, 2001] <http://www.edie.net/news/Archive/916.html>
55　Environment Agency, "Thames Water fined a record £250,000 for serious pollution offences at Sandcliff Road and at the Thames outfall, Erith, Kent," Press release, February 24, 2000.
56　Environment Agency, *Achieving the Quality, op. cit.*, pp. 1-2, 11, 23.
57　Byatt, *op. cit.*
58　Tabb, *op. cit.*, p. 29.
59　Drinking Water Inspectorate, "Letter from the Chief Inspector," and "Chief Inspector's Statement," *Drinking Water 2000*, July 11, 2001.
60　"Pipe dreams," *Economist*, January 4, 2001.
61　Ofwat, *Guaranteed Standards Scheme*, Information Note No. 4, May 1991, Revised February 2001.
62　Stuart Ogden and Fiona Anderson, "Representing Customers' Interests: The Case of the Privatized Water Industry in England and Wales," *Public Administration*, Vol. 73, Winter 1995, pp. 541-3, 550, 552, 555-6.
63　Ogden and Anderson, *op. cit.*, p. 536; Tabb, *op. cit.*, p. 20; and Office of Water Services, *Summary of the Director General's Annual Report 1997*, June 1998.
64　Alan Booker, Deputy Director General of Ofwat, "British privatization: balancing needs," *AWWA Journal*, March 1994, pp. 61-2.
65　David Wheeler and Rupesh Shah, *The UK Experience*, Paper and overheads prepared for Ontario SuperBuild Corporation Water Infrastructure Policy and Financing Forum, Toronto, December 2001, p. 8.
66　Canadian Union of Public Employees, "Privatization on our Doorstep," undated flyer regarding the private sector's possible involvement in treating Halifax's wastewater.

67   Canadian Union of Public Employees, "Privatization Facts," Water Watch campaign document, 1999.

68   Ontario Standing Committee on Resources Development, *Hansard*, April 14, 1997, 1050; and April 15, 1997, 1130.

69   Office of Water Services, "Unmeasured water" and "Unmeasured sewerage," unpublished figures provided by Andrea Lort, March 2, 2000.

70   "EA announces largest ever environmental improvement programme for waterways," Edie Weekly Summaries, July 7, 2000 [online] [consulted April 7, 2001] <http://www.edie.net/news/Archive/2913.html>

71   Burnell, *op. cit.*, p. 35; and Water UK, *op. cit.*, p. 46.

72   Office of Water Services, *Annual Report of the Director General of Water Services for the Period 1 April 1998 to 31 March 1999*, p. 25.

73   Byatt, *op. cit.*

74   Department of the Environment, Transport and the Regions, *Water Industry Act 1999: Delivering the Government's Objectives*, February 2000, p. 4.
Unsurprisingly, once the threat of disconnection was removed, increasing numbers of consumers failed to pay their water bills on time. By 2001, the number of late payers had risen by 10 percent, and more than 4.4 million households were in debt to water companies. (Matthew Jones, "Rise in number of customers failing to pay water bills," *Financial Times*, August 21, 2001.)

75   "Water bosses defend bonuses," BBC online network, June 28, 1998.

76   "Getting fat on water," BBC online network, March 23, 1998.

77   Byatt, *op. cit.*

78   Jamie Doward, "The cost of metering," *Observer*, June 14, 1998.

79   Institute for Fiscal Studies, "Taxing the Privatised Utilities," *The Economic Review*, Vol. 15, No. 4, April 1998.

80   Office of Water Services, *Paying By Meter*, Information Note No. 27, November 1995, Revised March 1996.

81   Richard Kingston, University of Leeds, citing National Consumer Council, *Credit and Debt: The Consumer Interest* (London: HMSO, 1990); and Office of Water Services, *Annual Report 1990*, 1991.

82   British Medical Association, Board of Science and Education, *Water: A Vital Resource*, 1994, p. 26.

83   Office of Water Services, "Water disconnections fall for the seventh year running," Press release, June 17, 1999.

84   Department of the Environment, Transport and the Regions, *Water Industry Act 1999: Delivering the Government's Objectives*, p. 4.

85   Canadian Environmental Law Association, "Keeping Municipal Water and Wastewater Services in Public Hands," undated draft resolution.

86   Ontario Standing Committee on Resources Development, *Hansard*, April 16, 1997, 1336, 1350.
87   Ontario Standing Committee on Resources Development, *Hansard*, April 16, 1997, 1030.
88   Ontario Standing Committee on Resources Development, *Hansard*, April 15, 1336.
89   Canadian Union of Public Employees, *Hostile Takeover: 1999 Annual Report on Privatization* (Ottawa: CUPE, January 1999), p. 21.
90   Judy Darcy, "Privatizing water leaves us high and dry," *Halifax Herald*, August 12, 1999.
91   Ontario Public Service Employees Union, *Focus on Water Privatization, op. cit.*
92   Ontario Standing Committee on Resources Development, *Hansard*, April 14, 1997, 940.
93   Privatization and the destruction of communities: What can we learn from the English experience? Event sponsored by the Ontario Public Service Employees Union as part of the *Jane Jacobs: Ideas that Matter* conference, Toronto, October 9, 1997.
94   Public Health Laboratory Service Communicable Disease Surveillance Centre NOIDS Database. Data provided by Douglas Harding, Information Officer, January 26 and 27, 2000.
95   British Medical Association, *op. cit.*, pp. 2, 12, 14, 39, 44.
96   *Ibid.*, pp. 13, 36, 37.
97   Alex Meredith, Letter to Elizabeth Brubaker, March 5, 1999.
98   Burnell, *op. cit.*, p. 48; and Water UK, *op. cit.*, p. 64.
99   "Water: Behind the Tide of Job Cuts," BBC online network, December 9, 1999.
100 Amalgamated Engineering and Electrical Union, *Employees' View of Privatization in the UK*, undated; and Danny Carrigan, E-mail to Elizabeth Brubaker, May 2, 2001.
101 Kathy Neal, Patrick J. Maloney, Jonas A. Marson, and Tamer E. Francis, *Restructuring America's Water Industry: Comparing Investor-Owned and Government-Owned Water Systems*, Policy Study No. 200 (Los Angeles: Reason Foundation, January 1996), p. 18.
102 Ontario Standing Committee on Resources Development, *Hansard*, April 15, 1997, 1336.
103 "Pipe dreams," *Economist*, January 4, 2001.
104 Michael Williamson, in *Water Delivery Systems: An International Comparison*, Transcript of panel discussion hosted by the Canadian Council for Public-Private Partnerships, November 1998, pp. 9, 29.
105 Byatt, *op. cit.*

CHAPTER FOUR

1   Craig Golding, *Water Utilities Privatization in France* (Toronto: Environment Probe, 1998), p. 68.

2   Jean-Louis Gazzaniga and Xavier Larrouy-Castera, "Eau potable et assainissement," *L'eau: usages et gestion* (Toulouse: Litec, Éditions du Juris-Classeur, August 1997), p. 5.

3   J. Elnaboulsi, *Organization, Management and Privatization in the French Water Industry*, report to the UN, ECLAC 4, 1998, forthcoming in *Environmental and Resources Economics*, pp. 3, 7, 9, 11.

4   Golding, *op. cit.*, pp. 31, 36; and Thompson Gow & Associates, *Canada's Untapped Resource: Public-Private Partnerships in Water Supply and Wastewater Treatment*, Prepared for Technology Transfer Office, Environmental Technology Advancement Directorate, Environment Canada, Technology Transfer Series 2E, September 1995, p. 18.
Le Code Permanent Environnement et Nuisances (Section 111, February 15, 1998, p. 2456) specifies that contracts should run no longer than 20 years without an extraordinary derogation.
According to Nicolas Spulber and Asghar Sabbaghi, private companies administered 65 percent of the country's wastewater treatment in 1995. *Economics of Water Resources: From Regulation to Privatization*, Second Edition (Boston: Kluwer Academic Publishers, 1998), p. 242.

5   Institute français de l'environnement, "Eau potable: diversité des services . . . grand écart des prix," *Les données de l'environnement*, No. 65, April, 2001, p. 2.

6   Elnaboulsi, *op. cit.*, p. 12.

7   Institut français de l'environnement, *Les indicateurs de performance environnemental de la France*, Extract from *Aménagement du territoire et environnement: Politiques et indicateurs*, July 2000, p. 183 [online] [consulted November 5, 2001] <http://www.ifen.fr/perf/perf2000/index.htm>

8   Organisation for Economic Co-operation and Development, *Environmental Performance Reviews: France* (Paris: OECD, 1997), p. 58.

9   Institut français de l'environnement, "Eaux continentales: La qualité," *Les chiffres-clés de l'environnement* [online] [consulted November 5, 2001] <http://www.ifen.fr/chifcle/qualite.htm>

10 Elnaboulsi, *op. cit.*, pp. 12-14.

11 Golding, *op. cit.*, pp. 18-20, 61-2.

12 Elnaboulsi, *op. cit.*, pp. 11, 15, 22; and Golding, *op. cit.*, pp. 69-71.

13 David Haarmeyer, *Privatizing Infrastructure: Options for Municipal Water-Supply Systems*, Policy Study No. 151 (Los Angeles: Reason Foundation, October 1992), p. 18.

14 Organisation for Economic Co-operation and Development, *op. cit.*, p. 146.

15 Blair T. Bower et al., *Incentives in Water Quality Management: France and the Ruhr Valley* (Washington: Johns Hopkins University Press, 1981), p. 107.

16 Réseau National des Données sur l'Eau, *L'Assainissement des Grands Villes* (Limoges: RNDE, 1998), p. 5.
The European Commission puts the agencies' investment in combatting sewage pollution between 1997 and 2001 at EUR 5 billion, or half of the amount budgeted for the schemes. European Commission, "Bathing Water Quality Annual Report, 2000 Bathing Season: France" [online] [consulted November 6, 2001] <http://europa.eu.int/water/water-bathing/report/fr.html>

17 Mikael Skou Andersen, *Governance by Green Taxes: Making Pollution Prevention Pay* (Manchester: Manchester University Press, 1994), p. 113.

18 Institute français de l'environnement, "Les dépenses des départements et régions en faveur de l'environnement," *Les données de l'environnement*, No. 49, November 1999, p. 2.

19 Ivan Chéret, "Managing Water: The French Model," *Valuing the Environment: Proceedings of the First Annual International Conference on Environmentally Sustainable Development*, eds. Ismail Serageldin and Andrew Steer (Washington: The World Bank, 1994), p. 85.

20 Institute français de l'environnement, "Les dépenses des départements et régions en faveur de l'environnement," *op. cit.*, p. 2.

21 Institut français de l'environnement, "La gestion des eaux usées 1990 - 1994," *Études et Travaux*, No. 10, September 1996, p. 41.

22 Institute français de l'environnement, "La dépense de protection de l'environnement en 1999: la reprise des investissements," *Les données de l'environnement*, No. 67, June 2001, p. 2.

23 On average, the price of water increased 66 percent between 1990 and 1997. Institute français de l'environnement, *L'environnement en France – Edition 1999 – Morceaux choisis* [online] [consulted November 5, 2001] <http://www.ifen/fr/choisis.htm>

24 Elnaboulsi, *op. cit.*, p. 16.

25 Fédération nationale dés collectivités concédantes et régies, *La Lettre S des services des eaux*, No. 103, April 16, 1997, p. 2.

26 Excerpts from report by la Cour des comptes, January 1997 [online] [consulted November 6, 2001] <http://www.marianne-en-ligne.fr/97-07-28/dessus-c.htm>

27 Ministère de l'Environnement, direction de l'eau, *Gestion des services d'eau et d'assainissement: les reformes en cours*, November 1995, p. 9.

28 "Water Music at Lyonnaise des Eaux in France," *Financial Times*, February 2, 1996 [online] [consulted November 6, 2001]

<http://www.transparency.org/documents/newsletter/96.1/Mar96.h
tml#France>

29 Henry Buller, "Privatization and Europeanization: The Changing
    Face of Water Supply in Britain and France," *Journal of Environmental
    Planning and Management*, Vol. 39, No. 4, 1996, p. 465.

30 "French Audit Court Lashes at Water Privatisation," *Privatisation
    News*, February 1997 [online] [consulted November 6, 2001]
    <http://www.transparency.org/documents/newsletter/97.2/NLJune9
    7-2.html#france>

31 Excerpts from report by la Cour des comptes, *op. cit.*

32 Henri-Benoit Loosdregt, Communication with Craig Golding, May
    1998.

33 Institut français de l'environnement, "Eaux continentales: La qual-
    ité," *op. cit.*

34 Institut français de l'environnement, *Les indicateurs de performance
    environnemental de la France, op. cit.*, pp. 176-7.

35 Institute français de l'environnement, "La préoccupation des
    Français pour la qualité de l'eau," *Les données de l'environnement*, No.
    57, August, 2000, p. 3.

36 Institut français de l'environnement, *Les indicateurs de performance
    environnemental de la France, op. cit.*, pp. 181-4.

37 European Commission, "Bathing Water Quality Annual Report,
    1998 Bathing Season: France."

38 Yves Coquin, Letter to Craig Golding, August 27, 1998.

39 European Commission, "Bathing Water Quality Annual Report,
    2000 Bathing Season: France" [online] [consulted November 6,
    2001] <http://europa.eu.int/water/water-bathing/report/fr.html>

40 Institut français de l'environnement, *Les indicateurs de performance
    environnemental de la France, op. cit.*, p. 108.

41 Réseau National des Données sur l'Eau, *L'Assainissement des Grands
    Villes, op. cit.*, p. 8.

42 Réseau National des Données sur l'Eau, *L'Assainissement des Grands
    Villes*, (Limoges: RNDE, 1997), pp. 10, 12, 14-5, 21-3; and Réseau
    National des Données sur l'Eau, *Mode d'exploitation des stations d'épu-
    ration des villes*, 1995.

One rough calculation of the average percentage of oxidisable material removed suggests that the big four water companies and the
public operators perform within two percentage points of each
other, with the latter performing slightly better than the former.
Because the calculation omits many communities for which data is
unavailable, is not weighted by population served or by volume of
pollution to be treated, and does not account for contract require-

ments or system characteristics, it is of very limited use.

43   Organisation for Economic Co-operation and Development, *op. cit.*, p. 156.

44   Jihad Elnaboulsi, *op. cit.*, p. 14.

45   Jean-Pierre Chevenement, Speech to Assemblée générale de l'association du corps préfectoral et des hauts fonctionnaires, November 1997.

46   *Le Code Permanent Environnement et Nuisances*, Section 238, February 15, 1998, p. 2494A.

CHAPTER FIVE

1   Thompson Gow and Associates, *Canada's Untapped Resource: Public-Private Partnerships in Water Supply and Wastewater Treatment*, Prepared for Technology Transfer Office, Environmental Technology Advancement Directorate, Environment Canada, Technology Transfer Series 2E, September 1995, pp. 1, 39.

2   Colin D'Cunha, Statement of Anticipated Evidence, the Walkerton Inquiry, p. 9; and "Education Days," Overheads Presented at the Walkerton Inquiry, June 28, 2001, pp. 7-8.

3   Ontario Ministry of the Environment, "Environment ministry completes inspection of 645 water treatment plants," News release, December 21, 2000.

4   Ingrid Peritz, "Quebec calls for urgent repair of drinking-water systems in 90 communities," *Globe and Mail*, August 19, 2000; and "Boil-water orders issued," *National Post*, August 19, 2000.

5   Sharon Boase, "One town in four now boiling water," *St. John's Telegram*, June 24, 2000.

6   British Columbia Provincial Health Officer, *Drinking Water Quality in British Columbia: The Public Health Perspective*, Provincial Health Officer's Annual Report 2000, October 2001, p.15.

7   "Saskatchewan warned of water threat," *National Post*, May 11, 2001. Part of the problem lies not in the water systems themselves but in the regulatory regimes governing them. In a national drinking water "report card" prompted by the Walkerton tragedy, the Sierra Legal Defence Fund graded each provincial or territorial regime by comparing requirements for source protection, testing, treatment, operator training, and public reporting to those in the U.S. Only four provinces – Alberta, Nova Scotia, Ontario, and Quebec – merited a B or B-. Manitoba, New Brunswick, the Northwest Territories, Nunavut, and Saskatchewan each rated a C or C-, while British

Columbia, Newfoundland, and the Yukon each received a D or D-. Prince Edward Island – which has no binding standards for testing or treatment, where disinfection is rare, where neither operators nor labs are certified, and where there are no binding requirements for notifying the public of problems – failed. (Randy Christensen and Ben Parfitt, *Waterproof: Canada's Drinking Water Report Card* (Vancouver: Sierra Legal Defence Fund, January 2001), pp. 37-40.)

8   Thompson Gow and Associates, *op. cit.*, p. 5.

9   Environment Canada, "Did you know?" *A Primer on Fresh Water: Questions and Answers*, 2000 [online] [consulted November 14, 2001] <http://www.ec.gc.ca/water/en/info/pubs/primer/e_prim11.htm>; and Jack Branswell, "Water crisis a matter of timing, critic says," *Toronto Star*, May 27, 2000.

10   Environment Canada, *Urban Water: Municipal Water Use and Wastewater Treatment*, SOE Bulletin No. 98-4, 1998.

11   Ontario Ministry of Environment and Energy, *Provincial Sewage and Water Inspection: Summary Report*, October 1992, pp. 20-1.

12   Doug Sider, former Associate Medical Officer of Health, Waterloo Region Community Health Department, E-mail to Environment Probe's Lisa Peryman, November 28, 2000.

13   United States Environmental Protection Agency, *Water On Tap: A Consumer's Guide to the Nation's Drinking Water*, EPA #815/K-97-002 (Washington: EPA, July 1997); "Milwaukee Crypto Settlement – $100k," *Public Works Financing*, Vol. 139, April 2000, p. 16; and Ajaib Singh, "An American Municipality Recovering from a Water Crisis," Presentation to *Safe and Clean Drinking Water Strategies*, Toronto, July 10, 2001.

14   Doug Saunders, "Water in 43 communities vulnerable to bug, minister says," *Globe and Mail*, March 30, 1996.

15   Susan Bourette, "Secret list cites 120 water woes," *Globe and Mail*, July 17, 2000; and Susan Bourette and Richard Mackie, "At least 48 communities issued boil-water order," *Globe and Mail*, July 19, 2000.

16   Martin Mittelstaedt, "Hazardous solvent found in Ontario groundwater," *Globe and Mail*, March 21, 2001.

17   Martin Mittelstaedt, "Ontario waterworks training falls short," *Globe and Mail*, December 26, 2000.

18   Alanna Mitchell, John Gray, and Rhéal Séguin, "Fear of farming," *Globe and Mail*, June 3, 2000.

19   Pierre Payment, E-mail to Environment Probe's Lisa Peryman, November 13, 2000; Pierre Payment *et al.*, "Occurrence of pathogenic microorganisms in the Saint Lawrence River (Canada) and comparison of health risks for populations using it as their source of

drinking water," *Canadian Journal of Microbiology* 46: 565-576; and Nicolas Van Praet, "Tap water under scrutiny," *Montreal Gazette*, April 27, 2000.

20 Andrew Orkin, "The message of Walkerton," Letter to the editor, *Globe and Mail*, May 30, 2000; and Mark MacKinnon, "Native reserves harbour dirty secrets," *Globe and Mail*, August 8, 2000.

21 Paul Taylor, "Illness flows from tap, lurks at grocer," *Globe and Mail*, April 14, 1997.

22 Canadian Press, "Parasite infects B.C. city's water supply," *Globe and Mail*, August 14, 1996; and Paul Taylor, "Illness flows from tap, lurks at grocer," *op. cit.*

23 British Columbia Provincial Health Officer, *op. cit.*, pp. 13-15.

24 "Water contains carcinogen," *National Post*, December 14, 1999.

25 "Problem drinking water," *National Post*, September 6, 1999; and "63 communities have contaminated water, tests show," *National Post*, January 11, 2000.

26 Environment Canada, *Urban Water, op. cit.*

27 Environment Canada, "Did you know?" *op. cit.*

28 Canadian Union of Public Employees, *Hostile Takeover: Annual Report on Privatization*, 1999, p. 23.

29 Miranda Holmes and Karen Wristen, *The National Sewage Report Card (Number Two)* (Vancouver: Sierra Legal Defence Fund, August 1999), p. 3.

30 Environment Canada, *Urban Water, op. cit.*

31 Martin Nantel, *Sewage Treatment and Disposal in Quebec: Environmental Effects* (Toronto: Environment Probe, 1995), p. 2.

32 Holmes and Wristen, *op. cit.*, p. 3.

33 Martin Nantel, *Troubled Waters: Municipal Wastewater Pollution on the Atlantic Coast* (Toronto: Environment Probe, 1996), p. 9.

34 Nantel, *Sewage Treatment and Disposal in Quebec, op. cit.*, pp. 5-7.

35 Ontario Ministry of the Environment, *1998 Waste Water Discharges Summary*, pp. 27-41; and Sierra Legal Defence Fund, *Who's Watching our Waters?*, May 2000, pp. 21-89.

36 British Columbia Ministry of Environment, Lands and Parks, "Latest Environmental Non-compliance Report Released," Press release, March 19, 1998.

37 Scott Simpson, "Victoria backs delays in cleaning toxic flow," *Vancouver Sun*, April 5, 2002.

38 British Columbia Ministry of Environment, Lands and Parks, "Environmental Non-compliance Report Released," Press release, November 24, 1999.

39 Nantel, *Troubled Waters, op. cit.*, p. 7.

40   Nantel, *Sewage Treatment and Disposal in Quebec, op. cit.*, p. 14.

41   National Round Table on the Environment and the Economy, *State of the Debate on the Environment and the Economy: Water and Wastewater Services in Canada*, 1996, pp. 10, 35.

42   Canadian Water and Wastewater Association, *Municipal Water and Wastewater Infrastructure: Estimated Investment Needs 1997 to 2012*, April 1997, revised April 1998, p. iv.

43   Thompson Gow and Associates, *op. cit.*, p. 6.

44   Office of the Provincial Auditor of Ontario, *1994 Annual Report*, p. 86; Ontario Standing Committee on Resources Development, *Hansard*, April 14, 1997, 1620; and April 15, 1997, 1120.

45   Nantel, *Sewage Treatment and Disposal in Quebec, op. cit.*, pp. 4-5; Nantel, *Troubled Waters, op. cit.*, pp. 9, 20; and Martin Nantel, *Municipal Wastewater Pollution in British Columbia* (Toronto: Environment Probe, 1996), p. 15.

46   National Round Table on the Environment and the Economy, *State of the Debate, op. cit.*, p. 8.

47   Ontario Standing Committee on Resources Development, *Hansard*, April 16, 1997, 1610.

48   Ontario Standing Committee on Resources Development, *Hansard*, April 15, 1997, 1510.

49   Michael Power, "Municipal Action Plan – Protecting Ontario's Water," Speech to the City of Windsor, June 12, 2000.

50   Ian Jack, "Sewer, Road Plans Drain Surplus," *National Post*, December 11, 2001.

51   Ontario Ministry of Environment and Energy, "Sterling establishes $200 million fund to protect quality of Ontario's drinking water," Press release, August 28, 1997; Ontario Ministry of Finance, "Ontario investing $2.1 billion in infrastructure," Press release, May 2, 2000; and British Columbia Ministry of Municipal Affairs, "Infrastructure funding emphasized in 1999 local government grants package," Press release, December 18, 1998.

52   Ontario Superbuild Corporation, "Investing in our Public Health and Safety Infrastructure," Backgrounder, August 10, 2000.

53   Price Waterhouse, *Study of Innovative Financing Approaches for Ontario Municipalities*, commissioned by the Municipal Finance Branch of the Ontario Ministry of Municipal Affairs, March 1991, pp. 13-4; Michael Shaen, ed., *The "3Ps" of Municipal Infrastructure: How Local Governments Can Use Public/Private Partnerships to Finance, Build and Operate Services*, Acumen Consulting Group, September 1997, p. 11; and Provincial-Municipal Investment Planning and Financing Mechanism Working Group, *Meeting the Challenge*, no date (the

group was established in November 1991), p. 18.

54 D. M. Tate, *Water Demand Management in Canada: A State-of-the-Art Review*, Social Science Series No. 23 (Ottawa: Environment Canada, 1990), p. 13; and Organisation for Economic Co-operation and Development, *The Price of Water: Trends in OECD Countries* (Paris: OECD, 1999), p. 79.

55 Thompson Gow and Associates, *op. cit.*, p. 13.

56 Steven Renzetti, "Municipal water supply and sewage treatment: costs, prices, and distortions," *Canadian Journal of Economics*, Vol. 32, No. 3, May 1999, p. 688.
    According to another estimate, Ontarians' charges typically cover 65 percent of the costs of water services. (Infrastructure Working Group Services Subcommittee, *GTA 2021 Infrastructure Requirements, Part 3: Water and Sewer Systems*, March 1992, p. 3.13.)

57 Canada, *Canada and Freshwater: Experience and Practices*, Monograph No. 6, Sustainable Development in Canada Monograph Series, 1998, p. 3.

58 Organisation for Economic Co-operation and Development, *OECD Economic Surveys: Canada 1999/2000*, September 2000, p. 130, citing Renzetti (1999).

59 Canadian Water and Wastewater Association, *op. cit.*, p. 31.

60 Thompson Gow and Associates, *op. cit.*, p. 8.

61 Commission sur la gestion de l'eau au Québec, "The commission opposes the privatization of municipal water services," Press release, May 3, 2000.

62 Thompson Gow and Associates, *op. cit.*, pp. v, 40, 41, 41.

63 National Round Table on the Environment and the Economy, *The Sustainable Cities Initiative: Putting the City at the Centre of Public-Private Infrastructure Investment*, no date, pp. 3-4.

64 National Round Table on the Environment and the Economy, *State of the Debate, op. cit.*, pp. 4, 17.

65 The Delta Partners, *Final Report: Awareness Workshops for Public-Private Partnerships in Wastewater Treatment in British Columbia*, March 1997.

66 Nova Scotia Department of Housing and Municipal Affairs, *Canada-Nova Scotia Cooperation Agreement to Promote Private Sector Participation in Municipal Infrastructure*, February 1997; and Grant Brennan, Telephone conversation with Elizabeth Brubaker, May 9, 2000.

67 Nova Scotia Department of Housing and Municipal Affairs, *Strategic Public-Private Partnering: A Guide for Nova Scotia Municipalities*, funded by the Canada-Nova Scotia Cooperation Agreement to Promote

Private Sector Participation in Municipal Infrastructure, no date, pp. 3-5.

68 Price Waterhouse, *op. cit.*, Appendix 1, pp. 2-3.

69 Provincial-Municipal Investment Planning and Financing Mechanism Working Group, *op. cit.*, pp. 20-1.

70 For a brief overview of many of these studies, see Elizabeth Brubaker, *The Privatization of Water Utilities: Supplementary Information*, Submission to the Walkerton Inquiry, August 13, 2001.

71 Ontario Cabinet Minute No. 7-29/96, Cabinet meeting of August 14, 1996, p. 1.

72 Larry Clay and Satish Dahr, Briefing note regarding MOEE Cabinet submission on the provincial role in sewer and water, August 6, 1996, p. 1.

73 Ontario Ministry of Environment, *1997 Business Planning Cycle Update*, September 1996, p. 15.

74 Submission to the Cabinet Committee on Privatization, May 1997, p. 11.

75 Ontario Ministry of Environment, *Short and Medium Term Municipal Capital Funding Scenarios: Explanatory Notes*, undated, p. 2.

76 Walkerton Inquiry, *Transcript*, June 27, 2001, p. 230, lines 3-10 and p. 235, lines 1-3.

77 Submission to the Cabinet Committee on Privatization, *op. cit.*, p. 7.

78 Ontario Standing Committee on Resources Development, *Hansard*, April 14, 1997, 950; and April 30, 1997, 1610.

79 Ontario Standing Committee on Resources Development, *Hansard*, April 16, 1997, 1030.

80 Ontario Standing Committee on Resources Development, *Hansard*, April 14, 1997, 940.

81 April Lindgren, "Businesses urged to bid for public services," *National Post*, February 10, 2000.

82 Martin Mittelstaedt, "Ontario to sell water, sewage agency," *Globe and Mail*, October 17, 1996; Daniel Girard, "Sterling says water sell-off in works," *Toronto Star*, October 18, 1996; "Ontario to sell clean water agency," *Financial Post*, October 18, 1996; and John Ibbitson, "Crown firm won bid, avoided selloff with contractual 'poison pill,'" *Globe and Mail*, April 6, 2000.

83 John Gray, "Tories study privatizing municipal water, sewage," *Globe and Mail*, June 13, 2000.

84 Richard Mackie, "Harris drops privatization; more E. coli found," *Globe and Mail*, June 15, 2000.

85 Richard Mackie, "Water systems to be revamped, Harris confirms," *Globe and Mail*, January 23, 2001.

86 Thompson Gow and Associates, *op. cit.*, p. 8.

87 Mylène Levac and Philip Wooldridge, "The fiscal impact of privatization in Canada," *Bank of Canada Review*, Summer 1997, p. 38.

88 Michael Klein, "The Good News and Bad News about the Cost of Capital," *Public Works Financing*, Vol. 138, March 2000, p. 25.

89 Karen Prokopec, *Framework for Alternative Financing: A Report by the Alternative Financing and Public- Private Partnerships Working Group*, September 5, 1997, pp. 1-2.

90 Canadian Council for Public-Private Partnerships, Letter to J. Fleming, Deputy Minister of Environment, February 27, 1998.

91 Eric Cunningham, Interview with Elizabeth Brubaker, February 2, 2001.

92 Guy Crittenden, "Closed for Business," *Report on Business Magazine*, June 1999, p. 56.

93 Corporate Planning Section, Finance & Administration, *Ontario Clean Water Agency – Overview of Mandate and Responsibilities*, March 2, 1998, p. 1.

94 Ontario Standing Committee on Resources Development, *Hansard*, April 16, 1997, 1610.

95 Guy Crittenden, *op. cit.*, p. 56.

96 Ontario Office of Privatization, *Review of the Ontario Clean Water Agency*, November, 1998, Volume 1 of 2, p. 47.

97 Ontario Standing Committee on Resources Development, *Hansard*, April 15, 1997, 1336, 1430, 1550. For other examples of similar concerns, see April 14, 1997, 940, 1050, 1400; April 15, 1997, 1140, 1540; and April 16, 1997, 1110, 1440.

98 Canadian Environmental Law Association, Backgrounder, May 29, 1998.

99 Eric Reguly, "Bulk water exports would wash jobs away," *Globe and Mail*, February 12, 2000.

100 Canadian Environmental Law Association, Canadian Union of Public Employees, and Council of Canadians, "Our Water is Not for Sale," Press release, December 7, 1998.

101 Steven Shrybman, *A Legal Opinion Concerning the Potential Impact of International Trade Disciplines on Proposals to Establish a Public-Private Partnership to Design Build and Operate a Water Filtration Plant in the Seymour Reservoir*, prepared for the Canadian Union of Public Employees, May 31, 2001; Canadian Union of Public Employees, "Trade deals threaten local decision-making – BC water service must remain public," Press release, June 4, 2001; and Canadian Union of Public Employees, "Vancouver water victory is huge," *Fastfacts*, July 5, 2001.

102 Peter Foster, "CUPE's water win is Canada's loss," Editorial, *National Post*, July 6, 2001.

103 Paul Lalonde, "Don't swallow CUPE's water arguments," *National Post*, August 13, 2001.

104 Peter Kirby, "Scary private water is an absurd myth," *National Post*, November 2, 2001.

105 Peter Kirby and David Doubilet, *Comments of Fasken Martineau DuMoulin LLP on the Shrybman Opinion*, Canadian Council for Public-Private Partnerships, Submission to the Walkerton Inquiry, September 19, 2001, pp. 2-3.

106 Canadian Union of Public Employees, *Hostile Takeover, op. cit.*, p. 23, citing November 1998 Vector Poll.

107 Claire Farid, John Jackson, and Karen Clark, *The Fate of the Great Lakes: Sustaining or Draining the Sweetwater Seas?* (Toronto: Canadian Environmental Law Assn. and Great Lakes United, February 1997), p. 69.

108 Canadian Council for Public-Private Partnerships, *Building Effective Partnerships*, September 1998, pp. 26, 40, 49, 50.

109 Canadian Council for Public-Private Partnerships, *Building Effective Partnerships, op. cit.*, pp. 4, 34, 39.

CHAPTER SIX

1 Canadian Council for Public-Private Partnerships, *Building Effective Partnerships*, September 1998, pp. 16-17.

2 Canadian Council for Public-Private Partnerships, *Public-Private Partnerships: Canadian Project and Activity Inventory – 1998*, September 1998, pp. 17, 54-5, 70.

3 Canadian Council for Public-Private Partnerships, *Public-Private Partnerships, op. cit.*, pp. 38, 54, 65, 39; and John Presta, Voice-mail to Elizabeth Brubaker, August 4, 2000.

4 Wally MacKinnon, E-mail to Elizabeth Brubaker, April 3, 2001.

5 United Water Canada, *Delegated Management*, undated brochure.

6 Dennis O'Connor, Report of the Walkerton Inquiry: Part Two (Toronto: Ontario Ministry of the Attorney General, 2002), p. 279.

7 Canadian Council for Public-Private Partnerships, *Public-Private Partnerships, op. cit.*, pp. 64-8, 74-5, 78-9.

8 USFilter, "Haldimand-Norfolk, Ontario," *Statement of Qualifications*, September 2000.

9 City of London, "Contract awarded for water supply systems," Press release, September 20, 2001.

10   "London, Ont., Water to Azurix, not OCWA," *Public Works Financing*, Vol. 154, September 2001, p. 11.

11   Larry McCabe, Fax to Elizabeth Brubaker, November 8, 2000.

12   Larry McCabe, E-mail to Elizabeth Brubaker, December 6, 2000.

13   USFilter, "Goderich, Ontario," *Statement of Qualifications*, September 2000.

14   Tim Cumming, "Town, USF sign deal to run water, wastewater," *Goderich Signal-Star*, November 1, 2000.

15   Delbert Shewfelt, Presentation to the Walkerton Inquiry's expert meeting on the implications of public and private operations on the safety of drinking water, July 5, 2001.

16   Canadian Council for Public-Private Partnerships, *Overview of Successful Public-Private Partnerships in the Water Sector*, November 2000, p. 13.

17   "Benefits of a Partnership Between the Town of Goderich and USF Canada," undated. Provided by Larry McCabe.

18   Christopher Guly, "Goderich water goes private," *Ottawa Citizen*, July 25, 2000.

19   Canadian Council for Public-Private Partnerships, *Public-Private Partnerships, op. cit.*, pp. 36-9; and Canadian Council for Public-Private Partnerships, *Overview of Successful Public-Private Partnerships, op. cit.*, pp. 6-7.

20   "Banff DBO sewage plant proposals," *Public Works Financing*, Vol. 148, February 2001, p. 5; and "Banff, Alberta, DBO to Earth Tech," *Public Works Financing*, Vol. 153, July-August 2001, p. 11.

21   Canadian Council for Public-Private Partnerships, *Public-Private Partnerships, op. cit.*, pp. 37-8; Michael Shaen, ed., *The "3Ps" of Municipal Infrastructure: How Local Governments Can Use Public/Private Partnerships to Finance, Build and Operate Services*, Acumen Consulting Group, September 1997, pp. 41-2; and Paul Waldie, "Private water system running in Alberta," *Financial Post*, February 12, 1994.

22   B.C. Ministry of Environment Lands and Parks, Water Management Branch, Utility Regulation Section, *Annual Report for the Period Ending March 31, 2000*, draft; and Rick Couroux, Telephone conversation with Elizabeth Brubaker, May 15, 2000.

23   Chester Merchant, Telephone conversation with Elizabeth Brubaker, May 17, 2000.

24   Mike Williams, Telephone conversation with Elizabeth Brubaker, May 10, 2000.

25   Greater Vancouver Regional District, Seymour Filtration: Project Overview, September 2000.

26   Larry Pynn, "GVRD cools to idea of private water plant," *Vancouver*

Sun, June 22, 2001.

27 Greater Vancouver Regional District, "GVWD decides against design-build-operate arrangement for construction of drinking water filtration facilities," Press release, June 29, 2001.

28 City of Kamloops, "Minutes of a regular meeting of the municipal council of the City of Kamloops," June 26, 2001, pp. 6-7.

29 City of Kamloops, "Minutes of a regular meeting of the municipal council of the City of Kamloops," July 10, 2001, pp. 2-6, 20-1.

30 Nova Scotia Department of Housing and Municipal Affairs, "Community Infrastructure Needs Assessment," July 1996.

31 Edward Greenspon, "Public works attracting private interests," *Globe and Mail*, January 8, 1994.

32 Halifax Harbour Solutions Symposium, "Background Summary," 1997.

33 K. R. Meech, "Request for Expressions of Interest, Harbour Solutions Project," Memo to Mayor Fitzgerald and members of Halifax Regional Council, July 2, 1998.

34 Marilla Stephenson, "HRM's next mega-project: shutting up the mayor," *Halifax Herald*, January 25, 2001; Chad Lucas, "City treads water on pace on harbour work," *Halifax Herald*, November 15, 2001; Jeffrey Simpson, "City picks cleanup company," *Halifax Herald*, December 6, 2001; and Jeffrey Simpson, "Harbour cleanup plan gets council approval," *Halifax Herald*, December 12, 2001.

35 Jeffrey Simpson, "Committee favours private bid for cleanup," *Halifax Herald*, October 30, 2001.

36 Halifax's estimates likely included land acquisition costs that do not appear in the consortium's budget and a "community integration fund" designed to help plants better fit into local areas. (Mike Kroger, Telephone conversation with Elizabeth Brubaker, May 14, 2002.)

37 Remi Couineau, Telephone conversation with Elizabeth Brubaker, May 14, 2002.

38 "Harbour gets green light," Editorial, *Halifax Herald*, December 8, 2001.

39 Richard Foot, "Moncton shuts down as broken water pipe leaves city high and dry," *National Post*, July 15, 1999.

40 City of Moncton, *Annual Water Quality Report*, 1998, pp. 3-6.

41 James Forster, "Free water today," *Moncton Times and Transcript*, October 1, 1999.

42 "City approves $23.1-M water treatment plant," *Moncton Times and Transcript*, April 7, 1998.

43 Charles Perry, "Water woes should be over next year," *Moncton Times and Transcript*, October 28, 1998.

44 City of Moncton, New Brunswick, and USFilter, "Canada's first major drinking water public-private partnership begins delivering water," Press release, October 21, 1999.
45 Christopher Guly, "Goderich water goes private," *Ottawa Citizen*, July 25, 2000.
46 City of Moncton, New Brunswick, and USFilter, *op. cit.*; L. E. Strang, E-mail to Elizabeth Brubaker, May 9, 2000; L. E. Strang, Letter to Ken Meech, Chief Administrative Officer for Halifax Regional Municipality, February 29, 2000; and USFilter, "Moncton, New Brunswick," *Statement of Qualifications*, September 2000.
47 "Clean water is worth cost," *Moncton Times and Transcript*, March 18, 1998.
48 James Forster, "Water treatment begins," *Moncton Times and Transcript*, October 21, 1999.
49 City of Moncton, *op. cit.*, pp. 4, 7.
50 James Forster, "Water treatment begins," *op. cit.*
51 USFilter, "Moncton, New Brunswick," *op. cit.*
52 L. E. Strang, "Getting the Process Right: Procurement for P3s," Presentation to *Narrowing the Gap*, Canadian Council for Public-Private Partnerships conference, November 28, 2000, Toronto.
53 Brian Murphy and L. E. Strang, Presentation to the Walkerton Inquiry's expert meeting on the financing of drinking water systems, June 20, 2001.
54 James Foster and Craig Babstock, "Moncton wants cost of fixing city water system," *Moncton Times and Transcript*, March 5, 2002.

CHAPTER SEVEN

1 On January 1, 2001, the six municipalities comprising the Region of Hamilton-Wentworth amalgamated to become the City of Hamilton. Employees and commentators tend to use the two names interchangeably.
2 Art Leitch, E-mail to editor@summitconnects.com, February 3, 2000.
3 Jon Wells, "Water deal with PUMC called a horror story," *Hamilton Spectator*, January 30, 1999.
4 Eric McGuinness, "Waste treatment facility will cost $11m," *Hamilton Spectator*, April 12, 2000.
5 Ken Peters, "Sludgegate probe finds $14m 'screw-up,'" *Hamilton Spectator*, July 30, 1992.
6 Jon Wells, "Return to tender," *Hamilton Spectator*, January 30, 1999.
7 Steve Buist, "The business of wastewater treatment," *Hamilton*

Spectator, November 28, 1998.

8   Michael Davison, "H-W workers cautioned on violence at work," *Hamilton Spectator*, December 4, 1992.

9   Michael Davison, "H-W stripper case angers politicians kept in the dark," *Hamilton Spectator*, April 21, 1992.

10 Davison, "H-W workers cautioned on violence at work," *op. cit.*

11 The Region of Hamilton-Wentworth, Philip Utilities Management Corporation, and Philip Environmental, *Plant Operations Agreement*, December 30, 1994, Article 12.01, pp. 56-8.

12 Buist, "The business of wastewater treatment," *op. cit.*

13 Nolan Bederman, "Philip Utilities Management Corp. and the Region of Hamilton-Wentworth: Public-Private Partnership for Wastewater Treatment," in *Case Studies in Public-Private Partnerships*, Nolan Bederman, Frank DeLuca, and Michael J. Trebilcock, jointly sponsored by the Canadian Council for Public-Private Partnerships and the Centre for the Study of State and Market at the University of Toronto Faculty of Law, August 1996, pp. 8, 12.

14 No author, "Cash may soon flow from sewers," *Hamilton Spectator*, April 15, 1994.

15 Editorial, "It reeks of potential: Water and sewage," *Hamilton Spectator*, April 22, 1994.

16 Jack MacDonald, "Sewage proposal is brass-ring chance to attract world market," *Hamilton Spectator*, April 27, 1994.

17 Mark McNeil, "Philip a link to Hamilton's hopes," *Hamilton Spectator*, March 30, 2001.

18 Regional Municipality of Hamilton-Wentworth, *Council News*, Volume 4, Issue 7, from the meeting of April 19, 1994.

19 John Anderson, *Privatising Water Treatment: The Hamilton Experience*, Report prepared for the Canadian Union of Public Employees, January 1999, p. 6, citing Leo Gohier's presentation at *Forming Successful Public Private Partnerships*, Toronto, May 1997.

20 Buist, "The business of wastewater treatment," *op. cit.*

21 Wells, "Return to tender," *op. cit.*

22 Terence Corcoran, "The Lost Temple of Honesty," *Globe and Mail*, May 3, 1996; and Bederman, *op. cit.*, p. 9.

23 Bederman, *op. cit.*, p. 12.

24 Buist, "The business of wastewater treatment," *op. cit.*

25 Bederman, *op. cit.*, p. 21.

26 *Ibid.*, p. 25.

27 *Ibid.*, pp. 9, 23-5, 29.

28 Ontario Standing Committee on Resources Development, *Hansard*, April 16, 1997,1600.

29 Wells, "Return to tender," *op. cit.*

30 *Plant Operations Agreement, op. cit.*, Article 5.01, p. 30, and Schedule M.

31 *Ibid.*, Article 5.01, pp. 35-8.

32 *Ibid.*, Article 11.01, pp. 53-4; Article 11.02, p. 56; Article 18.04, p. 71; Article 24.01, p. 74.

33 *Ibid.*, Article 6.03, p. 44; Article 6.04, pp. 44-5; Schedules F1, F2, F3.

34 The operator has since paid each employee $500 in a one-time buy-out of the profit-sharing provision in the contract. (Mark Hudson, Background on Region/PUMC/ANA, Memo to file (draft), December 19, 2000.)

35 Bederman, *op. cit.*, p. 41, citing comments made in a 1995 interview.

36 Philip Utilities Management Corp., *Water and Wastewater Operations and Maintenance*, April 12, 1999; and Jon Wells, "PUMC sold for $107m," *Hamilton Spectator*, March 27, 1999.

37 Jon Wells, "Seattle deal offers little consolation," *Hamilton Spectator*, February 1, 1999.

38 Editorial, "Region's haste on PUMC sale is unhealthy," *Hamilton Spectator*, May 4, 1999.

39 Wells, "Return to tender," *op. cit.*

40 Wells, "Seattle deal offers little consolation," *op. cit.*

41 Bederman, *op. cit.*, p. 29.

42 M. Stirrup and K. Stolch (signatories), *1998 Philip Utilities Management Corporation (PUMC) Performance Review (ENV99025)*, March 25, 1999.

43 *Plant Operations Agreement: Fourth Amending Agreement and Consent*, Section 2 (d), May 17, 1999, p. 6 and Schedule S.

44 Azurix maintains that PUMC fulfilled its obligations regarding an international training centre "as reasonably as possible." The centre, it maintains, was set up and ran for three years before closing down due to a lack of students. (Hudson, Background on Region/PUMC/ANA, *op. cit.*)

45 Stirrup and Stolch, *op. cit.*

46 *Plant Operations Agreement: Fourth Amending Agreement and Consent*, Section 2 (d), May 17, 1999, p. 6.

47 Stuart Smith, E-mail to Elizabeth Brubaker, December 9, 2000.

48 John Stokes, *An International Water Business Headquartered in Hamilton*, Presentation to the City of Hamilton Council, June 20, 2001. The avoided expenditures likely included the Woodward Avenue pre-treatment facility, the closed incinerator, the computerized automation system, and centrifuges.

49 Eric McGuinness, "Azurix sale could place city in financial squeeze,"

Hamilton Spectator, October 20, 2001.

50   W. Murray McCulloch, "Second thoughts needed on plan for private operation," *Hamilton Spectator*, June 16, 1994.

51   Louise Knox, *Remedial Action Plan for Hamilton Harbour: 1998 Status Report*, September 1998, pp. iv, 2.

52   City of Hamilton, Spreadsheet containing data on the monitoring and posting of Hamilton area beaches, 1995- 2000.

53   Ken Peters, "Don't blame us, Philip president says of flood," *Hamilton Spectator*, September 7, 1996.

54   Mark McNeil, "Alarm at sewage plant went unheeded," *Hamilton Spectator*, February 19, 1997.

55   Ken Peters, "Charges pending over wastewater plant flood," *Hamilton Spectator*, September 12, 1996.

56   Mark McNeil, "Is Ontario going soft on polluters?" *Hamilton Spectator*, February 6, 1998.

57   Ontario Standing Committee on Resources Development, *Hansard*, April 16, 1997, 1610.

58   McNeil, "Is Ontario going soft on polluters?" *op. cit.*

59   Anderson, *op. cit.*, p. 12.

60   Wells, "Return to tender," *op. cit.*

61   Steve Buist, "No investigation at Woodward Avenue plant," *Hamilton Spectator*, November 28, 1998.

62   Steve Buist, "Ministry investigates Woodward sewage spill," *Hamilton Spectator*, February 2, 1999; and Wylie Rogers, "Illegal wastewater dumping triples," *Hamilton Spectator*, October 11, 1999. Confirmed by effluent quality data provided by Water Quality Manager Mark Stirrup, April 12, 2000.

63   Steve Buist, "Wastewater plant put on notice," *Hamilton Spectator*, December 23, 1999. Confirmed by effluent quality data provided by Water Quality Manager Mark Stirrup, April 12, 2000.

64   Buist, "Ministry investigates Woodward sewage spill," *op. cit.*

65   Eric McGuinness, "Hamilton's sewage plant spawns stinky concerns," *Hamilton Spectator*, July 23, 2001.

66   George Sorger, "The sad state of our sewers," *Hamilton Spectator*, October 14, 1999.

67   George Sorger, "We must demand higher sewage standards," Letter to the editor, *Hamilton Spectator*, February 12, 1999.

68   Buist, "Ministry investigates Woodward sewage spill," *op. cit.*; Steve Buist, "Sewage plant can't handle volume," *Hamilton Spectator*, February 2, 1999; and Peter Yemen, "Take back control of our essential services," Letter to the editor, *Hamilton Spectator*, February 3, 1999.

69 John Percy, Letter to Greg Hoath, March 18, 1999.
70 Rick Hughes, "Ministry, region at odds over effluent measuring," *Hamilton Spectator*, March 13, 1999.
71 Buist, "No investigation at Woodward Avenue plant," *op. cit.*
72 No author, "Sewage spill has mayor fuming," *Hamilton Spectator*, January 12, 1999.
73 Lisa Hepfner, "Sewage flows into creek bed," *Hamilton Spectator*, September 1, 1999.
74 Buist, "Sewage plant can't handle volume," *op. cit.*
75 Buist, "No investigation at Woodward Avenue plant," *op. cit.*
76 *Ibid.*
77 Rick Hughes, "Region gives good review for PUMC," *Hamilton Spectator*, April 14, 1999.
78 Wells, "Return to tender," *op. cit.*
79 Calculations based on effluent quality data for the period from 1990 through 1999 provided by Water Quality Manager Mark Stirrup, April 12, 2000.
80 John Stokes and Mark Hudson, Meeting with Elizabeth Brubaker, February 9, 2001.
81 Leo Gohier and Jeff McIntyre, Meeting with Elizabeth Brubaker, December 20, 2000; and McGuinness, "Hamilton's sewage plant spawns stinky concerns," *op. cit.*
82 *Plant Operations Agreement, op. cit.*, Article 3.04, pp. 10-11.
83 *Ibid.*, Article 3.02, pp. 5-6, Schedule B, pp. 84-5.
84 Anderson, *op. cit.*, p. 7.
85 Eric McGuinness, "Sewage upgrade tab is $570m," *Hamilton Spectator*, April 28, 2000.
    One 2001 report on the state of Hamilton's infrastructure indicated that the sewage treatment plant needs $600 million in upgrades and may take 30 years to fix. According to the report, the city also needs to spend $2 billion over 10 years to repair sewer and water pipes, a quarter of which are between 80 and 100 years old. (Rick Hughes, "City must spend $2b on sewers," *Hamilton Spectator*, February 20, 2001.) Another report put the cost of Hamilton's wastewater infrastructure projects at $630 million. (City of Hamilton and Azurix North America, *Achieving the Vision: The True Story of Hamilton and Outsourcing*, Submission to the Walkerton Inquiry, August 2001, p. 7.)
86 Buist, "The business of wastewater treatment," *op. cit.*
87 Hughes, "Region gives good review for PUMC," *op. cit.*
88 *Ibid.*
89 Fred Eisenberger, "You can hold me accountable," Interview with editorial board, *Hamilton Spectator*, October 18, 2000.

90 Greg Hoath, Letter to John Percy, November 19, 1998; Greg Hoath, Letter to John Percy, February 23, 1999; and Greg Hoath, Letter to John Percy, March 30, 1999.

91 John Percy, Letter to Greg Hoath, December 3, 1998; and John Percy, Letter to Greg Hoath, March 18, 1999.

92 Buist, "No investigation at Woodward Avenue plant," *op. cit.*

93 Buist, "Wastewater plant put on notice," *op. cit.*

94 Terry O'Neill, Telephone conversation with Elizabeth Brubaker, January 26, 2001. Thirteen of the charges are described in Eric McGuinness, "Sewage plant firm charged," *Hamilton Spectator*, January 16, 2001.

95 Lisa Hepfner, "Sewage plant operator cited for pollution, licence offences," *Hamilton Spectator*, July 7, 2001; and Mark Hudson, Telephone conversation with Elizabeth Brubaker, July 2001.

96 City of Hamilton and Azurix North America, *Achieving the Vision, op. cit.*, p. 7; and Mark Hudson, Telephone conversation with Elizabeth Brubaker, November 30, 2001.

97 Eric McGuinness, "Guilty plea on sewage charges," *Hamilton Spectator*, August 3, 2001; and Ontario Ministry of the Environment, "Azurix fined a total of $184 for environmental violations," Press release, November 8, 2001.

98 Regional Municipality of Hamilton-Wentworth, *Council News, op. cit.*

99 Mr. Gohier attributes these more limited savings to Hamilton's having run one of the lowest-cost operations in North America, leaving less room for savings. (Leo Gohier and Jeff McIntyre, Meeting with Elizabeth Brubaker, December 20, 2000.)

100 *Plant Operations Agreement, op. cit.*, Article 5.01, p. 38; and Leo Gohier, E-mail to Environment Probe's Kevin Lacey, April 19, 2000.

101 Mark Hudson, Background on Region/PUMC/ANA, *op. cit.*

102 Stuart Smith, E-mail to Elizabeth Brubaker, December 9, 2000.

103 *Plant Operations Agreement, op. cit.*, Article 5.01, pp. 30-1, 33-4.

104 *Ibid.*, Article 3.02, pp. 6-7, Article 3.04, p. 11, Article 3.06, pp. 12-3, Article 3.10, pp. 14-5, Article 3.12, p. 16, Article 3.13, p. 16, Article 3.22, p. 21, Article 3.23, p. 22, Article 4.02, p. 23, Article 4.04, p. 25, Article 4.06, p. 27, Article 5.01, p. 34, Article 5.04, p. 40, Article 5.05, p. 41, Article 6.08, p. 47, Schedule B, p. 84. One such adjustment concerned the difference in hydro costs resulting from a change in operations requested by the region after the 1996 sewage spill. The parties' agreement to use a three-year trend to calculate the difference contributed to the delay in finalizing costs. (Leo Gohier and Jeff McIntyre, Meeting with Elizabeth Brubaker, December 20, 2000.)

105 Stokes, *An International Water Business, op. cit.*

106 City of Hamilton and Azurix North America, *Achieving the Vision, op. cit.*, p. 10.
107 City of Hamilton, Proposed Sale of Azurix North America Corp. to American Water Services, Inc., Schedule "D" Term Sheet, pp. 15-16; and Dave Clancy, E-mail to Elizabeth Brubaker, February 19, 2002.
108 Eisenberger, "You can hold me accountable," *op. cit.*
109 Richard Dooley and Brian Flinn, "A terrible mistake," *Halifax Daily News*, February 15, 2000.
110 Steve Arnold, "Philip water-sewer deal works: Cooke," *Hamilton Spectator*, September 22, 1998.
111 Steve Arnold, "Councillors red-faced over sewage surprises," *Hamilton Spectator*, April 10, 1996.
112 Steve Arnold, "20 losing jobs at sewage plant," *Hamilton Spectator*, March 23, 1996.
113 Mark Hudson, E-mails to Elizabeth Brubaker, February 12, February 16, and July 4, 2001.
Others' figures on staffing levels vary. The *Plant Operations Agreement* (Schedules G1, G2, and G3) lists 138 positions (not all of which were filled) in 1994. According to Leo Gohier and Jeff McIntyre, approximately 25 voluntary retirements at the end of 1994 brought the number of positions down to approximately 113, approximately 12 of which were vacant. Thus, PUMC started out with approximately 101 filled positions. (Meeting with Elizabeth Brubaker, December 20, 2000.) According to Greg Hoath, the number of filled, unionized positions fell from 108 before privatization to 32 in January 2001. (E-mails to Elizabeth Brubaker, January 25 and February 21, 2001.)
David Cameron reports that the numbers fell from 130 (including 121 bargaining-unit employees) on January 1, 1995, to 47 (including 31 in bargaining unit) in June 2001. (David Cameron, *The Relationship Between Different Ownership and Management Regimes and Drinking Water Safety*, Discussion paper prepared for the Walkerton Inquiry, amended July 4, 2001, p. 104.)
114 Greg Hoath, Telephone conversations with Elizabeth Brubaker, January 22 and 29, 2001; and City of Hamilton and Azurix North America, *Achieving the Vision, op. cit.*, p. 13.
115 City of Hamilton and Azurix North America, *Achieving the Vision, op. cit.*, p. 9.
116 Leo Gohier and Jeff McIntyre, Meeting with Elizabeth Brubaker, December 20, 2000.
117 Bederman, *op. cit.*, pp. 44-5.
118 Stirrup and Stolch, *op. cit.*
119 City of Hamilton and Azurix North America, *Achieving the Vision, op. cit.*, p. 9.

120 Eric McGuinness, "Region scraps \$2.1m sewage presses," *Hamilton Spectator*, January 10, 2000.

121 Bederman, *op. cit.* pp. 46-8; Ken Peters, "Region, Philip have \$25-million spat," *Hamilton Spectator*, October 11, 1995; no author, "Region, Philip to steam into court," *Hamilton Spectator*, October 18, 1995; and Jim Poling, "Philip, region agree on using SWARU steam," *Hamilton Spectator*, November 18, 1995.

122 Leo Gohier, E-mail to Elizabeth Brubaker, November 22, 2000.

123 Anderson, *op. cit.*, p. 9; and Mark McNeil, "Rough ride for Philip experiment," *Hamilton Spectator*, September 14, 1996. Mr. Gohier minimizes the severity of these disagreements, calling them "issues" and "growing pains" and pointing out that the parties have agreed on a resolution process to be used in the event of disputes. (Leo Gohier, E-mail to Elizabeth Brubaker, November 22, 2000; and Leo Gohier and Jeff McIntyre, Meeting with Elizabeth Brubaker, December 20, 2000.)

124 Robert Crane, *Performance Appraisal of Philip Utilities Management Corporation, January 1, 1995 to December 31, 1995*, March 1996, pp. 3-4. Reproduced in Anderson, Appendix 1.

125 Bruce Livesey, "What's in Toronto's drinking water?" *Eye*, July 6, 2000.

126 Greg Hoath, Telephone conversation with Elizabeth Brubaker, January 22, 2001; and Cameron, *op. cit.*, pp. 104, 106-7.

127 Salim Loxley and John Loxley, *An Analysis of a Public-Private Sector-Partnership: The Hamilton-Wentworth-Philip Utilities Management Corporation PPP*, Prepared for the Canadian Union of Public Employees, September 1999, pp. 11, 15, 22.

128 Eric McGuinness, "Water problems blamed on efforts to save money," *Hamilton Spectator*, July 13, 1999; and Greg Hoath, Telephone conversation with Elizabeth Brubaker, January 22, 2001.

129 Mark McNeil, "Sewers are starting to fall apart," *Hamilton Spectator*, December 7, 1996.

130 *Plant Operations Agreement: Third Amending Agreement*, Section 13, January 1, 1997, pp. 8-9.

131 *Ibid.*, Article 4.05, pp. 25-6.

132 Anderson, *op. cit.*, p. 8; Greg Hoath, Telephone conversation with Elizabeth Brubaker, January 22, 2001; Greg Hoath, Speech to Hamilton-Wentworth Taxpayers Coalition, April 15, 1999, pp. 2-3; and Cameron, *op. cit.*, p. 107.

133 Wells, "Water deal with PUMC called a horror story," *op. cit.*

134 Leo Gohier, E-mail to Environment Probe's Lisa Peryman, November 10, 2000.

135 Greg Hoath, Telephone conversation with Elizabeth Brubaker, January 22, 2001.

136 *Plant Operations Agreement, op. cit.*, Article 3.25, p. 22.

137 *Ibid.*, Article 12.01, p. 58.

138 Anderson, *op. cit.*, Appendix A, Citing memorandum from Neil Everson, Manager of Business Development, Region of Hamilton-Wentworth Economic Development Department, March 7, 1996.

139 Andrew Dreschel, "Pot calling the kettle black," *Hamilton Spectator*, May 25, 2001.

140 Anderson, *op. cit.*, pp. 14-5.

141 *Plant Operations Agreement, op. cit.*, Article 3.07, p. 13, Article 5.07, p. 41.

142 *Ibid.*, Article 8.01, p. 52.

143 Loxley and Loxley, *op. cit.*, p. 15.

144 McNeil, "Rough ride for Philip experiment," *op. cit.*

145 Wells, "Return to tender," *op. cit.*

146 Dooley and Flinn, *op. cit.*

147 Gord McNulty, "Municipal water change no call for rash decisions," editorial, *Hamilton Spectator*, August 9, 2001.

148 Bederman, *op. cit.*, pp. 25, 51.

149 Wells, "Return to tender," *op. cit.*

150 Eric McGuinness, "Philip sale deal squeaks through on 14-13 vote," *Hamilton Spectator*, May 5, 1999; Dooley and Flinn, *op. cit.*; and Loxley and Loxley, *op. cit.* p. 24.

151 No author, "Firms chase region's water service contract," *Hamilton Spectator*, April 28, 1999.

152 Jon Wells, "Latest water system deal raises eyebrows," *Hamilton Spectator*, May 13, 1999.

153 Editorial, "Region's haste on PUMC sale is unhealthy," *Hamilton Spectator*, May 4, 1999.

154 Eric McGuinness, "How U.S. suitor won a chance to buy PUMC," *Hamilton Spectator*, May 6, 1999.

155 *Plant Operations Agreement: Fourth Amending Agreement and Consent*, Sections 2(b), 2(c), 2(e) pp. 4-7 and Schedule Q; and Leitch, *op. cit.*

156 McGuinness, "Philip sale deal squeaks through on 14-13 vote," *op. cit.*

157 John Munro, "I know how the programs work," Interview with editorial board, *Hamilton Spectator*, October 28, 2000.

158 Wells, "Latest water system deal raises eyebrows," *op. cit.*

159 Wells, "Return to tender," *op. cit.*

160 Negotiating the original agreement took Hamilton less than seven months and cost it less than $100,000. According to Leo Gohier,

other jurisdictions – Atlanta, Milwaukee, and Peel – spent one-and-a half to two years and $1.5 million to $2 million negotiating their privatizations. (Leo Gohier and Jeff McIntyre, Meeting with Elizabeth Brubaker, December 20, 2000.)

161 Anderson, *op. cit.*, p. 21.

162 The Milwaukee Metropolitan Sewerage District took several measures to reduce conflicts regarding maintenance and capital costs. It carefully defined the terms; attached to the contract a schedule of 46 items that would fall under one category or the other; and required the private operator to pay the first US$5,000 of each capital expenditure as an incentive to perform adequate maintenance. (Anne Spray Kinney, "Milwaukee's O&M Contract Protects Its Public Assets," *Public Works Financing*, Vol. 152, June 2001, p. 1.)

163 Anderson, *op. cit.*, pp. 16, 23.

164 Dooley and Flinn, *op. cit.*

CHAPTER EIGHT

1 The Canadian Council for Public-Private Partnerships, *Building Effective Partnerships* (Toronto: CCPPP, September 1998), pp. 26, 10, 13.

2 Robert W. Poole Jr., "The Limits of Privatization," in *Privatization: Tactics and Techniques*, ed. Michael Walker (Vancouver: The Fraser Institute, 1988), p. 94.

3 William Eggers, "The Nuts and Bolts: Overcoming the Obstacles to Privatization," *Commonwealth Conversations* (Harrisburg, PA: The Commonwealth Foundation for Public Policy Alternatives, April 1997), p. 2.

4 Patrick McManus, "Design Competition, Fair Margins Help Municipalities," *Public Works Financing*, Vol. 135, December 1999, p. 18.

5 Robin A. Johnson and Norman Walzer, "City Officials Seek More Information on Privatization," *Illinois Municipal Review*, April 1997, p. 7.

6 Robin A. Johnson and Norman Walzer, "Effects of Privatization on Municipal Employees in Illinois," *Illinois Municipal Review*, December 1996, p. 13; and Johnson and Walzer, "City Officials Seek More Information on Privatization," *op. cit.*, p. 7. The survey, conducted in 1995-96, was sponsored by the Illinois Municipal League, the Illinois Institute for Rural Affairs at Western Illinois University, and the state's Local Government Affairs Division.

7   Canadian Union of Public Employees, *Hostile Takeover: 1999 Annual Report on Privatization*, January 1999, p. 3.

8   Canadian Union of Public Employees, *Who's Pushing Privatization? 2000 Annual Report on Privatization*, February 2000, p. 4.

9   Canadian Union of Public Employees, *The Real Story Behind 'PPPs,'* undated.

10 Canadian Union of Public Employees, *Hostile Takeover, op. cit.,* pp. 33, 40.

11 American Federation of State, County and Municipal Employees, "Members Cite Privatization as Biggest Threat to Their Jobs," *AFSCME Leader,* June 1999.

12 American Federation of State, County and Municipal Employees, "Privatization: Pros and Cons," undated.

13 Public Services Privatisation Research Unit, *The Privatization Network* (London: PSIRU, January 1996).

14 Canadian Union of Public Employees, *Fighting Privatization: Update,* March 1999.

15 Despite CUPE's willingness to challenge new facilities in which it has no existing stake, some believe that it is easier to establish new private facilities than it is to privatize existing public facilities, since the former do not yet have a workforce with a vested interest in public operations. As one municipal representative told the Canadian Council for Public-Private Partnerships, "We are very closely watched by the unions here and have to tread very carefully in developing partnerships. That's one of the reasons most plans look at new ventures rather than contracting our city jobs." (The Canadian Council for Public-Private Partnerships, *Building Effective Partnerships, op. cit.,* p. 25.)

16 Canadian Council for Public-Private Partnerships, *Water Delivery Systems: An International Comparison* (Toronto: CCPPP, November 1998), p. 27.

17 See Chapter 7, note 113.

18 Theo Marks, E-mails to Elizabeth Brubaker, May 23 and 26, 2001.

19 "IRS Rules in Favor of Pension Portability," *Public Works Financing,* Vol. 129, May 1999, p. 4.

20 Eggers, *op. cit.,* p. 5.

21 The National Commission for Employment Policy, "The Long Term Employment Implications of Privatization: Evidence from Selected U.S. Cities and Counties" (Washington, D.C.: NCEP, March 1989). Cited by the Reason Foundation in "Political and Organizational Strategies for Streamlining," *Privatization Database* [online] [consulted December 2, 2001] <http://privati-

zation.org/Collection/PracticesAndStrategies/Political_and_Orga
nizational_Strategies.html>

22 The Reason Foundation, *op. cit.*

23 Johnson and Walzer, "Effects of Privatization," *op. cit.*, p. 13.

24 Eggers, *op. cit.*, p. 5.

25 Madsen Pirie, "Principles of Privatization," in *Privatization: Tactics and Techniques, op. cit.*, p. 21.

26 Economic Developers Council of Ontario, "Provincial Legislation and Public/Private Partnerships," *EDCO Exchange*, Vol. 1, No. 24, October 27, 1999.

27 Poole, "The Limits of Privatization," *op. cit.*, p. 95.

28 United States General Accounting Office, *Privatization: Lessons Learned by State and Local Governments*, Report to the Chairman, House Republican Task Force on Privatization, March 1997, p. 16.

29 American Federation of State, County and Municipal Employees, "Members Cite Privatization," *op. cit.*

30 Canadian Council for Public-Private Partnerships, *Water Delivery Systems, op. cit.*, p. 27.

31 "Public Bidders Run Well in Ohio Water," *Public Works Financing*, Vol. 134, November 1999, p. 8; "Scranton, Pa., O&M Sole-Sourced" and "Birmingham, Ala., ConOps Bids," *Public Works Financing*, Vol. 127, March 1999, pp. 8, 10; Milwaukee Metropolitan Sewerage District, *Competitive Contract Annual Report*, March 1999, p. 4; American Federation of State, County, and Municipal Employees, "No layoffs for 20 years," *AFSCME Leader*, June 1999; and "Four Vie for Springfield, Mass., Ops," *Public Works Financing*, Vol. 131, July-August 1999, p. 12.

32 Adrian Moore, "Long-term Partnerships in Water and Sewage Utilities: Economic, Political, and Policy Implications," in Universities Council on Water Resources, *Issue No. 117: Private Sector Participation in Urban Water Supply*, October 2000, p. 22.

33 "U.S. Water Wins Stonington, Conn.," *Public Works Financing*, Vol. 132, September 1999, p. 16.

34 Charmagne Helton, "Focus on Sewer Privatization," *Atlanta Journal / Constitution*, April 1, 1997.

35 *Amended and Restated Agreement for the Operation and Maintenance of the City of Indianapolis, Indiana, Advanced Wastewater Treatment Facilities*, Contract between the City of Indianapolis and White River Environmental Partnership, Amended December 1997, p. 33.

36 Danny Carrigan, *Employees' View of Privatization in the UK* (Bromley, Kent: Amalgamated Engineering and Electrical Union, 2001), p. 4; and Danny Carrigan, E-mail to Elizabeth Brubaker, May 2, 2001.

37 Canadian Union of Public Employees, *Hostile Takeover, op. cit.*, p. 34.

38 Johnson and Walzer, "Effects of Privatization," *op. cit.*, p. 14.

39 Dale Belman and John Heywood, *The Truth About Public Employees: Underpaid or Overpaid?* (Washington, D.C.: Economic Policy Institute, April 1993), pp. 5, 13.

40 James Rusk, "Council Orders Wage Increase," *Globe and Mail*, March 2, 2000.

41 National Center for Policy Analysis, "Living Wage," *Executive Alert*, Vol. 13, No. 6, November/December 1999, p. 1.

42 Paul Kengor and Grant Gulibon, *"Poison Pills" for Privatization: Legislative Attempts at Regulating Competitive Contracting* (Pittsburgh: Allegheny Institute for Public Policy, December 1996), p. 6.

43 United States General Accounting Office, *op. cit.*, p. 14.

44 Kengor and Gulibon, *op. cit.*, p. 7.

45 John O'Leary and William D. Eggers, *Privatization and Public Employees: Guidelines for Fair Treatment*, How-To Guide No. 9 (Los Angeles: Reason Foundation, September 1993).

46 Kengor and Gulibon, *op. cit.*, p. 7.

47 Rusk, *op. cit.*

48 Eggers, *op. cit.*, p. 3. Another source puts the number of privatizations during the two years before the law at 20 and says that two privatizations occurred in the two years following the law. (United States General Accounting Office, *op. cit.*, p. 14.)

49 Canadian Council for Public-Private Partnerships, *Water Delivery Systems, op. cit.*, p. 14.

50 *Amended and Restated Agreement, op. cit.*, p. 33.

51 Canadian Council for Public-Private Partnerships, *Water Delivery Systems, op. cit.*, p. 14.

52 Milwaukee Metropolitan Sewerage District, *Competitive Contract Annual Report*, March 1999, p. 4; and Canadian Council for Public-Private Partnerships, *Water Delivery Systems, op. cit.*, p. 14.

53 "IRS Rules in Favor of Pension Portability," *op. cit.*, p. 1.

54 Robert W. Poole Jr., "Privatization for Economic Development," in *The Privatization Process: A Worldwide Perspective*, ed. Terry L. Anderson and Peter J Hill (Lanham, Maryland: Rowman & Littlefield, 1996), pp. 7-8; and William L. Megginson, Robert C. Nash, and Matthias van Randenborgh, "The Financial and Operating Performance of Newly Privatized Firms: An International Empirical Analysis," in *The Privatization Process, ibid.*, p. 117.

55 Peter Young, *The Lessons of Privatization* (Washington: Center for International Private Enterprise, 1998), pp. 4-5.

56 Pirie, "The Principles and Practice of Privatization," *Fraser Forum*,

Special Issue, May 1987, pp. 9, 12; Pirie, *Policymaking and Privatization: Ten Lessons from Experience* (Washington: Center for International Private Enterprise, 1998), p. 3; and Kathy Neal, Patrick J. Maloney, Jonas A. Marson, and Tamer E. Francis, *Restructuring America's Water Industry: Comparing Investor-Owned and Government-Owned Water Systems*, Policy Study No. 200 (Los Angeles: Reason Foundation, January 1996), p. 18.

57 Bruce Little, "Union Membership Shrinking in Every Corner of the Economy," *Globe and Mail*, September 6, 1999.

58 Johnson and Walzer, "Effects of Privatization," *op. cit.*, p. 13.

59 "IRS Rules in Favor of Pension Portability," *op. cit.*, pp. 3-4.

60 White River Environmental Partnership, *City of Indianapolis Contract Operations of the AWT Facilities and Collection System: Fifth Year Summary of Activities*, p. 2.

61 Milwaukee Metropolitan Sewerage District, *Competitive Contracting Annual Report*, March 2000. Discrepancy in figures corrected in Mark Kass, E-mail to to Elizabeth Brubaker, April 13, 2000.

62 Eggers, *op. cit.*, p. 3.

63 Florencio López-de-Silanes, Andrei Shleifer, and Robert W. Vishny, "Privatization in the United States," *RAND Journal of Economics*, Vol. 28, No. 3, Autumn 1997, p. 468.

64 Michael L. Mills and Charles D. VanEaton, "Privatization Objections Are Weak," *Policy Fax Perspective* (Palatine, Illinois: The Heartland Institute, July 1992), p. 1.

65 Michael Walker, ed., *Privatization: Tactics and Techniques* (Vancouver: Fraser Institute, 1988), p. 265.

66 Wally MacKinnon, E-mail to Elizabeth Brubaker, April 20, 2001.

67 Stephen Goldsmith, *The Twenty-First Century City* (Washington: Regnery, 1997), p. 55.

68 White River Environmental Partnership, *op. cit.*, p. 2.

69 A Glass Half Full or a Glass Half Empty: Unions and Business Leaders and their Thoughts on the Benefits of Public-Private Partnerships, *Narrowing the Gap*, The Canadian Council for Public-Private Partnerships's Eighth Annual Conference, Toronto, November 27, 2000.

70 Joseph Mancinelli, "Presentation to the Canadian Council for Public-Private Partnerships," November 27, 2000, pp. 4, 7, 8.

71 Greg Hoath, Telephone conversation with Elizabeth Brubaker, January 29, 2001.

72 Pirie, "Principles of Privatization," *op. cit.*, pp. 6-9; and Pirie, *Policymaking and Privatization, op. cit.* p. 3.

73 Canadian Union of Public Employees, *Hostile Takeover, op. cit.*, pp. 59-60.

74   The Canadian Council for Public-Private Partnerships, *Building Effective Partnerships, op. cit.*, p. 24.

CHAPTER NINE

1   United States Environmental Protection Agency, "Monterey County Water Suppliers Face Drinking Water Penalties," Region 9 news release, September 13, 2000; and United States District Court for the Northern District of California, San Jose Division, Order granting plaintiff's motions for partial summary judgment in *United States of America v. Alisal Water Corporation et al.*, August 23, 2000.
2   United States Environmental Protection Agency, "Florida Wastewater Treatment Facility and Owner Sentenced," Headquarters press release R-174, November 21, 2000.
3   Cat Lazaroff, "Puerto Rico Resort Fined for Dumping Sewage Into Sea," *Environment News Service*, December 29, 2000; and "Puerto Rico Resort Fined $430,000 for Sewage Discharges," *Environment News Service's Ameriscan*, September 14, 2001.
4   "When a man knows he is to be hanged in a fortnight, it concentrates his mind wonderfully." From James Boswell, *The Life of Samuel Johnson*, quoting a letter from Johnson to Boswell dated September 19, 1777.
5   Michael Trebilcock, testifying as an expert at *Energy Probe et al. v. The Attorney General of Canada*, October 14, 1993, Transcript Volume 3, 283, 27-30.
6   Cited in Allen M. Linden, *Canadian Tort Law*, Sixth Edition (Toronto: Butterworths, 1997), p. 7.
7   Experience in diverse fields confirms that strict liability – under which non-negligent defendants are held liable – increases incentives for responsible behaviour. Stricter product liability laws in the United States have led to the improved safety of many products. Similarly, increased medical malpractice premiums have prompted doctors to change their procedures. (Michael Trebilcock and Ralph Winter, *The Impact of the Nuclear Liability Act on Safety Incentives in the Nuclear Power Industry*, Exhibit 967 in *Energy Probe et al. v. The Attorney General of Canada*, April 29, 1993, pp. 9-12, 22-5.) For more on the deterrent value of liability, see Linden, *ibid*, pp. 7-13, 97.
8   Les Leyne, "Worst offenders get off easiest on polluters' list," *Victoria Times Colonist*, October 1, 1997.
9   Martin Mittelstaedt, "Water polluters escaping prosecution," *Globe and Mail*, March 1, 1999.

10 Canadian Environmental Defence Fund, "Enforcement," Distributed at Walkerton Inquiry meeting, Toronto, May 23, 2001.

11 Jamie Benidickson, *The Development of Water Supply and Sewage Infrastructure in Ontario, 1880-1990s: Legal and Institutional Aspects of Public Health and Environmental History*, Background paper prepared for the Walkerton Inquiry, Draft, February 2001.
For a detailed examination of Ontario's failure to enforce the laws and regulations governing water and wastewater utilities, see Elizabeth Brubaker, *The Promise of Privatization*, Study prepared for the Walkerton Inquiry on behalf of Energy Probe Research Foundation, April 2001, pp. 51-69.

12 In its review of international best practices for environmental compliance assurance, the Executive Resource Group stressed that cooperative approaches to abatement only work when backed up by the credible threat of coercive enforcement action. It cited a 1996 survey of corporate environmental managers, conducted by KPMG, that found that companies implement best environmental management practices because they have a legal duty to comply with regulations and are concerned about the potential for Board of Directors liability. (Executive Resource Group, "Environmental Compliance Assurance," *Managing the Environment*, Vol. 2, Research Paper 1, December 2000, pp. 8-10.)
The Energy Probe Research Foundation has proposed distancing abatement from enforcement by moving the latter from the Ministry of the Environment to the Ministry of the Solicitor General. Such a move, it argues, would help depoliticize enforcement. It would reduce conflicts between abatement and enforcement and transform enforcement into an independent, straightforward, policing function. (Thomas Adams, *Guiding and Controlling Ontario's Future Water and Wastewater Services*, Study prepared for the Walkerton Inquiry on behalf of Energy Probe Research Foundation, April 2001, pp. 31-3.)

13 F.A. Hayek, *The Road to Serfdom*, 1944, cited by Stephen Littlechild, *Privatisation, Competition and Regulation*, Wincott Memorial Lecture, delivered October 14, 1999, IEA Occasional Paper 110 (London: Institute of Economic Affairs, 2000), p. 4.

14 David Kinnersley, *Coming Clean: The Politics of Water and the Environment* (London: Penguin Books, 1994), pp. 47, 53; and Rt. Hon. Nicholas Ridley, Speech, May 22, 1987, Reprinted by the National Rivers Authority, *The Government's Proposals for a Public Regulatory Body in a Privatised Water Industry*, December 1997, p. 41.

15 Kinnersley, *ibid.* p. 204.

16 Lord Crickhowell, *House of Lords Hansard*, April 17, 1989, col. 579.

17 David Wallen, "Profiteering claims mark British privatization," *Globe and Mail*, October 17, 1996.

18 Environment Agency, "Environmental Gains Marred by Pollution Failures," Press release, July 26, 2000.

19 Ontario Standing Committee on Resources Development, *Hansard*, April 16, 1997, 1600.

20 Anne Spray Kinney, "Milwaukee's O&M Contract Protects Its Public Assets," *Public Works Financing*, Vol. 152, June 2001, p. 2.

21 Walkerton Inquiry, *Transcript*, June 27, 2001, p. 233, line 11 - p. 234, line 8.

22 Mervin Daub, *Regulation of Private Enterprise Vs Direct Control of Crown Corporations: A Comparison of Gas and Electricity in Ontario* (Toronto: Energy Probe, 1992), pp. 3, 8.

23 Jerry Ellig, *The $7.7 Billion Mistake: Federal Barriers to State and Local Privatization*, Joint Economic Committee Staff Report, February 1996, p. 1.

24 Robert W. Poole, Jr., "The Limits of Privatization," in *Privatization: Tactics and Techniques*, Proceedings of an International Symposium, ed. Michael A. Walker (Vancouver: The Fraser Institute, 1988), p. 96.

25 Robert W. Poole, Jr., Speech to The Canadian Council for Public-Private Partnerships, Toronto, May 18, 1999.

26 Bill Eggers, "The Nuts and Bolts: Overcoming the Obstacles to Privatization," Speech published as part of the "Commonwealth Conversations" series by the Commonwealth Foundation for Public Policy Alternatives, April 1997, p. 7.

27 Kathy Neal, Patrick J. Maloney, Jonas A. Marson, and Tamer E. Francis, *Restructuring America's Water Industry: Comparing Investor-Owned and Government-Owned Water Systems*, Policy Study No. 200 (Los Angeles: Reason Foundation, January 1996), p. 17.

28 Insurance, properly managed, will not reduce the deterrent of liability. If an insurance company monitors risks and charges higher premiums to reflect higher risks, the utility's incentive to avoid risk remains intact. As Justice Allen Linden explains in the context of automobile insurance, "Whereas liability insurance dulls the financial incentive of *drivers* to use care, it sharpens the motivation of insurance *companies* to prevent accidents. . . . Insurers can stimulate their insureds to institute safer practices." (Linden, *op. cit.*, pp. 10-11.) In a debate on liability in the pages of *Public Works Financing*, two authors note that "insurers' willingness to provide coverage is a strong indicator of a contractor's ability to do the job as specified. . . . In effect, the surety does much of the due diligence for the public sector

partner. Given its experience and the fact that the surety is taking substantial risk and has much at stake in the capabilities (financial and technical) of the service provider, it probably can be viewed as doing a better job of due diligence than the public sector partner." (Douglas Herbst and David Mackenzie, "Unlimited, Irrevocable, Unreasonable Liability," *Public Works Financing*, Vol. 147, January 2001, pp. 29-30.)

29   Conference Board of Canada, Letter and brochure promoting *Managing Reputation and Brands and Strategic Assets*, February 2001.

30   John Stokes, Meeting with Elizabeth Brubaker and Mark Hudson, February 9, 2001.

31   Linden, *op. cit.*, p. 28.

32   For more information on Crown immunity, see Duncan W. Glaholt and Markus Rotterdam, *Construction Contracting for Public Entities: Claims Against the Crown*, Glaholt & Associates, August 2000; and Evert Van Woudenberg, *Recent Cases on the Liability of Municipalities*, Published as part of a Gardiner, Roberts seminar on municipal liability, November 7, 1996.

33   Government of Ontario, *Proceedings Against the Crown Act*, Section 14.

34   Kirk M. Baert, "Class Actions and Other Legal Considerations in Cases of Environmental and Other Mass Disasters," Paper delivered at *Guaranteeing Safe and Clean Drinking Water for Your Community*, Strategy Institute conference, Toronto, September 27-29, 2000, pp. 52, 73.

35   *Regina v. Eldorado Nuclear Limited* (1984) 8 CCC (3rd) 449 (SCC) at p. 456.

36   William Bishop, "The Rational Strength of the Private Law Model," (1990), *40 University of Toronto Law Journal*, pp. 663-9 at p. 668. Also see *Anns v. Merton London Borough Council*, [1978] A. C. 728.

37   Bishop, *ibid.*, p. 664.

38   *City of Kamloops v. Nielsen*, [1984] 2 SCR 2, at pp. 9-10.

39   *Just v. British Columbia* (1989), 64 D.L.R. (4th) 689 (S.C.C.), at pp. 704-6.

40   *Brown v. British Columbia (Ministry of Transportation and Highways)* [1994] 1 S.C.R. 420 at pp. 441-2, 444.

41   *Oosthoek v. Thunder Bay (City)* (1995), 24 M.P.L.R. (2d) 25.

42   Government of Ontario, *Municipal Act*, Section 331.

43   Van Woudenberg, *op. cit.*, pp. 7, 13.

44   Government of Ontario, *Municipal Act*, Section 50.

45   David Cohen, "Suing the State," (1990), 40 *University of Toronto Law Journal*, pp. 630-62, at p. 661.

46   Bishop, *op. cit.*, p. 664.

47   The following paragraphs draw on Elizabeth Brubaker, *Property Rights*

in the *Defence of Nature* (London: Earthscan, 1995), pp. 93-112.

48 *Tock et al. v. St. John's Metropolitan Area Board* (1989), 64 D.L.R. (4th) 620, pp. 645-8.

49 For a detailed discussion, see Brubaker, *Property Rights in the Defence of Nature, op. cit.*, pp. 83-92.

50 *Burgess v. The City of Woodstock*, [1955] O.R. 814; and *Stephens v. The Village of Richmond Hill*, [1955] O.R. 806, aff'd [1956] O.R. 88.

51 Government of Ontario, *Ontario Water Resources Act*, Section 59.

52 Richard L. Stroup and Roger E. Meiners, *Cutting Green Tape: Toxic Pollutants, Environmental Regulation and the Law* (Rutgers: Transaction Publishers, 2000), p. 14.

53 Experience in other fields confirms that immunizing people or industries from risk and responsibility decreases their level of care. After Quebec adopted a no-fault automobile insurance system in 1978, automobile fatalities rose; Australia's no-fault scheme similarly increased fatalities. Likewise, industries that, thanks to government regulation, do not bear the costs of environmental destruction are unlikely to invest adequately in systems that preserve clean air, land, and water. (Trebilcock and Winter, *op. cit.*, pp. 10, 22.)

54 Jihad Elnaboulsi discussed regulators' informational disadvantages in *The Optimal Economic Regulation and the French Experience of Privatization*, Prepared for American Water Works Association, 1997 Annual Conference, Atlanta, pp. 19-21, 29.

55 Trebilcock and Winter, *op. cit.*, pp. 39-40.

56 Joel Franklin Brenner, "Nuisance Law and the Industrial Revolution," *Journal of Legal Studies* 3, no. 2 (June 1974), pp. 403-33 at p. 423. One act establishing sewage works is described in *Pride of Derby and Derbyshire Angling Association Ld. and Another v. British Celanese Ld. and Others*, [1953] 1 Ch. 149. The 1901 Derby Corporation Act, while establishing sewage disposal works, had specifically prohibited nuisances: "The sewage disposal works constructed . . . shall at all times hereafter be conducted so that the same shall not be a nuisance and in particular the corporation shall not allow any noxious or offensive effluvia to escape therefrom or do or permit or suffer any other act which shall be a nuisance or injurious to the health or reasonable comfort of the inhabitants of Spondon."

57 Jo-Christy Brown and Roger E. Meiners, *Common Law Approaches to Pollution and Toxic Tort Litigation*, in Stroup and Meiners, *op. cit.*, pp. 99-127, at p. 119.

58 For information on the use of trespass, nuisance, and riparian rights to prevent water pollution in Canada, see Brubaker, *Property Rights in the Defence of Nature, op. cit.*, pp. 29-67, 223-91. For information on

the traditional reliance on tort law to protect the US environment, and the extent to which statute law has preempted tort law in the last 30 years, see Bruce Yandle, "Common-Law Protection of Environmental Rights," *Common Sense and Common Law for the Environment: Creating Wealth in Hummingbird Economies* (Lanham: Rowman and Littlefield, 1997), pp. 87-118; and Brown and Meiners, *op. cit.*

59   *Crowther v. Town of Cobourg* (1912), 1 D.L.R. 40 (Ont. H.C.).

60   *Ibid.* at p. 43, citing *Roberts v. Gwyrfai District Council*, [1899] 2 Ch. D. 608.

61   Steven Shavell, "Liability for Harm Versus Regulation of Safety," *The Journal of Legal Studies*, Volume 13, June 1984, pp. 357-74; and Guido Calabresi and Jon T. Hirschoff, "Toward a Test for Strict Liability in Torts," *The Yale Law Journal*, Volume 81, Number 6, May 1972, pp. 1055- 1085.

62   Daniel K. Benjamin, "Contracting for Health and Safety: Risk Perception and Rational Choice," in Stroup and Meiners, *op. cit.*, pp. 223-49, at pp. 244-5.

63   Shavell, *op. cit.*, p. 365.

64   Dan Westell, "Epcor/OCWA Eye Seymour Water DBO," *Public Works Financing*, Vol. 145, November 2000, p. 24.

65   David Estrin and John Swaigen, *Environment on Trial: A Guide to Ontario Environmental Law and Policy*, Third Edition (Toronto: Emond Montgomery Publications, 1993), p. 79.

CHAPTER TEN

1   See, for example, S. C. Littlechild, *Economic Regulation of Privatised Water Authorities*, Report submitted to the Department of the Environment (London: Her Majesty's Stationery Office, 1986), p. 5; David Haarmeyer, *Privatizing Infrastructure: Options for Municipal Water-Supply Systems*, Policy Study No. 151 (Los Angeles: Reason Foundation, October 1992), p. 4; and J. Elnaboulsi, *Organization, Management and Privatization in the French Water Industry*, Unpublished report to the United Nations Economic Commission for Latin America and the Caribbean, 1997, p. 5.

2   Michael Klein, "IWaPs [Independent Water Providers] at the Gate,"*Public Works Financing*, Vol. 132, September 1999, p. 23.

3   Nicolas Spulber and Asghar Sabbaghi, *Economics of Water Resources: From Regulation to Privatization* (Boston: Kluwer Academic Publishers, Second Edition, 1998), pp. 96, 209-10.

Penelope Brook Cowen and Tyler Cowen point out that competing pipe systems occurred in Canada and the United States in the 19th century, and can still be found in Hong Kong, where pipes supply both sea-water for flushing and treated fresh water for other uses. (Penelope Brook Cowen and Tyler Cowen, "Deregulated Private Water Supply: A Policy Option for Developing Countries," Final draft, for publication in the *Cato Journal*, February 9, 1998, p. 6.)

4   Ofwat, *The Current State of Market Competition*, July 2000, pp. 2-3; and Ofwat, *Market Competition in the Water and Sewerage Industry*, Information Note No. 10, April 1992, Revised August 2000.

5   Alan Booker, Deputy Director General of Ofwat, "British privatization: balancing needs," *AWWA Journal*, March 1994, p. 58; Philip Fletcher, Director General of Water Services, "The Future Agenda of Competition in Water," Speech to the Economist Conference, October 10, 2000; Ofwat, *The Current State of Market Competition, op. cit.*, p. 2; and Ofwat, *Market Competition in the Water and Sewerage Industry, op. cit.*

6   Fletcher, "The Future Agenda of Competition in Water," *op. cit.*; and Philip Fletcher, "Regulatory Developments: Moving Towards Total Competition for Utilities?," Speech to *The Future of Utilities*, The Adam Smith Institute's 6th Annual Conference, London, March 14, 2001.

7   "Wales' Largest Water User Changes Supplier," Enviro-Logic On-Line, undated. [online] [consulted May 30, 2001] <http://www.enviro-logic.com/Content.asp?WCI=ShottonP>; and "New 'inset' water company wins first major supply contract," Edie Weekly Summaries, May 7, 1999. [online] [consulted May 22, 2001] <http://www.edie.net/news/Archive/1133.html>

8   Fletcher, "The Future Agenda of Competition in Water," *op. cit.*

9   Ofwat, *What is the Competition Act 1998?*, Information Note No. 45, May 2000.

10 Holly Brown, Information Services Officer, Ofwat, E-mail to Elizabeth Brubaker, June 6, 2001.

11 In July 2001, the Department for Environment, Food and Rural Affairs announced its plans to facilitate the trading of water abstraction licences.

12 Fletcher, "Regulatory Developments: Moving Towards Total Competition for Utilities?," *op. cit.*

13 Littlechild, *Economic Regulation of Privatised Water Authorities, op. cit.*, pp. 1-2, 11.

14 Ofwat, *Annual Report of the Director General of Water Services For the Period 1 April 1998 to 31 March 1999*, p. 22.

15 Littlechild, *Economic Regulation of Privatised Water Authorities, op. cit.,* pp. 8, 26.

16 Professor Littlechild is not alone in his criticism of rate-or-return regulation. In their seminal 1962 article, Harvey Averch and Leland Johnson identified the inefficiency and excessive investment that may result from rate-of-return regulation of the telephone and telegraph industry. (Harvey Averch and Leland Johnson, "Behavior of the Firm Under Regulatory Constraint," *American Economic Review,* Vol. 52, No. 5, December 1962, pp. 1052-69.) Regarding rate-of-return regulation of water utilities in particular, David Haarmeyer, among others, has charged the system with impeding innovation and has recommended replacing it with price-cap regulation. (Haarmeyer, *Privatizing Infrastructure: Options for Municipal Water-Supply Systems, op. cit,* pp. 1-2, 25, 30; and David Haarmeyer, "Privatizing Infrastructure: Options for Municipal Systems," *AWWA Journal,* March 1994, p. 48.) A policy study for the Reason Foundation attributed the U.S. water industry's inability to compete internationally to the inefficiencies inherent in an industry regulated by a rate-of-return framework. (Kathy Neal, Patrick J. Maloney, Jonas A. Marson, and Tamer E. Francis, *Restructuring America's Water Industry: Comparing Investor-Owned and Government-Owned Water Systems,* Policy Study No. 200 (Los Angeles: Reason Foundation, January 1996), pp. 15-16.) Tyler Cowen has likewise accused rate-of-return regulation in the United States of fostering water companies that are "sluggish, overstaffed, and bureaucratic." (Tyler Cowen, "Three Principles for Sound Water Policy," Speech to Institute of Economic Affairs Water Conference, November 1997.)

17 Littlechild, *Economic Regulation of Privatised Water Authorities, op. cit.,* pp. 1-3, 11, 26-30.
Ofwat's Director of Costs and Performance explains that, over 30 years, customers will realize 68 percent of the value of operating cost savings and water companies will realize 32 percent. (W. H. Emery, Letter to Jane Peatch, July 18, 2001.)

18 Stephen Littlechild, *Privatisation, Competition and Regulation,* Wincott Memorial Lecture, delivered October 14, 1999, IEA Occasional Paper 110, London: Institute of Economic Affairs, 2000, pp. 7-8, citing Israel M. Kirzner, *The Perils of Regulation: A Market-Process Approach* (Coral Gables: University of Miami School of Law, Law and Economics Center, 1978), p. 16.

19 In the five years following privatization, Ks varied widely among companies. For the new water and sewage companies, they ranged from 1.9 percent to 19 percent. For the old water-only companies,

they ranged from -11.2 percent to 22.5 percent. (Ofwat, *Price Limits (K) 1990-91 - 2001-02*, undated.)

20 Booker, *op. cit.*, pp. 58-9.

21 David S. Saal and David Parker, "Productivity and Price Performance in the Privatized Water and Sewerage Companies of England and Wales," *Journal of Regulatory Economics* 20:1, 2001, p. 87.

22 "Who regulates the regulators?" Editorial, *Financial Times*, May 20, 1996.

23 "UK Water: From Riches to Rags," *Public Works Financing*, Vol. 142, July/August 2000, p. 2.

Although estimates of profits and returns vary, all suggest that the companies and their shareholders did very well. The *Telegraph* estimates that after privatization, the water and sewage companies' profits climbed from £1.3 billion in 1990 to £2.1 billion in 1997. (Mary Fagan, "Byatt opens floodgates to competition," *Electronic Telegraph*, August 23, 1998.) Jean Shaoul calculates that dividends to parent companies amounted to £5.63 billion in the first four years of private ownership – a sum that exceeded the original purchase price of £5.25 billion. (Jean Shaoul, "A Critical Financial Analysis of the Performance of Privatised Industries: The Case of the Water Industry in England and Wales," *Critical Perspectives on Accounting* (1997) 8, p. 499.)

According to Ian Byatt, Director General of Water Services, shareholders' returns amounted to between 11 and 16 percent annually in real terms in the decade following privatization. ("Life after the 1999 Periodic Review," Speech to Royal Aeronautical Society, January 26, 2000.) An Institute for Fiscal Studies paper reports that, in the first four years after privatization, share prices rose 93 percent more than share-price growth for a broad index of companies. (Lucy Chennells, "The Windfall Tax," *Fiscal Studies*, Vol. 18, no. 3, 1997, pp. 280-1.)

24 Institute for Fiscal Studies, "Windfall tax on the privatised utilities," *The Economic Review*, Vol. 15, no. 4, April 1998.

25 "Pipe dreams," *The Economist*, January 4, 2001.

Public Works Financing called the 1999 price caps "confiscatory." Noting that water companies' share values fell by an average of 40 percent in the year following the new price caps, it quoted one utility advisor as saying, "Investors have really been clobbered." ("UK Water Regulator Rejects Kelda Exit Strategy," *Public Works Financing*, Vol. 142, July/August 2000, pp. 1-2.)

26 City Editor, "Kelda finds an income stream in a dried-up backwater," *Electronic Telegraph*, June 15, 2000.

27   "Ofwat refuses Kelda's proposals to restructure Yorkshire water," Edie Weekly Summaries, July 28, 2000. [on- line] [consulted July 28, 2001] <http://www.edie.net/news/Archive/3015.html>

28   "Water regulator allows non-profit company to attempt to buy Welsh Water," Edie Weekly Summaries, February 2, 2001. [on-line] [consulted February 3, 2001] <http://www.edie.net/news/Archive/3784.html> World Bank economist Michael Klein notes that debt financing would weaken capital market discipline and that the tax advantages of the new not-for-profit owners would reallocate risks from shareholders to taxpayers. He warns, "The squeeze put on by the regulator de facto pushes private companies to consider options that start looking suspiciously like the first step towards renationalization." (Michael Klein, "Privatization – Nationalization: The Wheel Keeps Turning," Public Works Financing, Vol. 145, November 2000, p. 22.)

29   Ofwat, The Role of the Regulator, undated. [online] [consulted May 22, 2001] <http://www.ofwat.gov.uk/rolereg.htm>

30   Ibid.

31   In a discussion of "the conundrum of price regulation," the World Bank's Michael Klein warned that regulators that pursue consumer interests at the expense of producer interests risk lowering prices to the point where they are insufficient to cover costs, which include adequate shareholder returns. The resulting under-investment, operating problems, and decline in the quality of service may prompt renationalization. (Klein, "Privatization – Nationalization: The Wheel Keeps Turning," op. cit.)

32   Janice A. Beecher, G. Richard Dreese, and John D. Stanford, Regulatory Implications of Water and Wastewater Utility Privatization (Columbus: National Regulatory Research Institute, July 1995), p. v.

33   G. Richard Dreese and Janice A. Beecher, "To Privatize Or Not to Privatize: A POTW question for the 1990s," Water Environment and Technology, January 1997, p. 52.

34   Ontario Standing Committee on Resources Development, Hansard, April 16, 1997, 1600.

35   Neal et al., op. cit., p. 16.

36   Beecher et al., op. cit., p. 127.

37   Ibid., pp. 127-8.

38   G. O. Barron, Letter to Elizabeth Brubaker, June 27, 2001.

39   Beecher et al., op. cit., pp. 124, 133, 149.

40   Thomas Adams, Guiding and Controlling Ontario's Future Water and Wastewater Services: User Pay and Full Cost Pricing, Independent Economic Regulation, and Strengthened Environmental Law Enforcement, Submission on behalf of Energy Probe Research Foundation to the

Walkerton Inquiry, April 2001, pp. 20-22.

41 Janice A. Beecher, "The Rationale for Regulating Municipal Contract Services," *Public Works Financing*, Vol. 145, November 2000, pp. 18-20.

42 For a detailed picture of what a regulatory system should look like, see Adams, *op. cit.*, pp. 8-24.

43 Beecher et al., *op. cit.*, pp. 142, 147, 149.

44 *Ibid.*, pp. 146, 150.

45 William G. Reinhardt, "U.S. Water/Wastewater Privatization Company Given Highest Award for Performance, Quality," *Public Works Financing*, Vol. 145, November 2000, p. 2.

46 The Canadian Council for Public-Private Partnerships, *Public-Private Partnerships: Canadian Project and Activity Inventory — 1998* (Toronto: CCPPP, September 1998), pp. 77, 122.

47 Doug Paisley and Elizabeth Brubaker, *The Ontario Clean Water Agency: Supplementary Information*, Submission on behalf of Energy Probe Research Foundation to the Walkerton Inquiry, September 2001, pp. 9-10.

CONCLUSION

1 "Privatized Operation and Managed Competition," *Public Works*, Vol. 128, No. 6, May 1997, p. 85; and Bureau of Governmental Research, *Privatization of Water and Wastewater Systems in New Orleans* (New Orleans: BGR, June 2001), p. 16 and Appendix E.

2 The idea of allowing public departments and their unions to compete for work against private firms has been called "a significant public-sector trend." (Amy Shanker and Len Rodman, "Public-private partnerships,"*Journal AWWA*, April 1996, p. 106.) Of the six governments interviewed by the U.S. General Accounting Office, four permitted – and in some cases encouraged by providing training and consultants – at least some employee groups to compete with private-sector bidders. (United States General Accounting Office, *Privatization: Lessons Learned by State and Local Governments*, Report to the Chairman, House Republican Task Force on Privatization, GAO/GGD-97-48, March 1997, pp. 6, 34-5.) Workers commonly bid on contracts in Indianapolis, where Mayor Goldsmith long emphasized competition over privatization. Of the 66 services that were put out to tender by 1996, 37 – generally the larger services – were awarded to private firms. Unions won 20 of the 29 services they bid on, while some were split between the two sectors. (Jeff Bowden, Glena Carr and Judi Storrer, *New directions in municipal services: Competitive contracting and*

alternative service delivery in North American municipalities (Toronto: ICURR Press, 1997), p. 6.)

3   When Cincinnati Water Works bid to operate the water and waste-water systems in neighbouring Clermont County, city council authorized it to offer the same performance and surety bonds required of private bidders, with city ratepayers assuming the risk. ("Ohio County Picks Private Water Bid," *Public Works Financing*, Vol.139, April 2000, p. 10.) In New Orleans, in contrast, the draft MOUs for private and public bidders differed, since Sewage and Water Board staff could not legally assume financial liability for errors or default. ("New Orleans to Offer First Water + Sewer," *Public Works Financing*, Vol. 148, February 2001, pp. 9-10.) In either case, it is the public, rather than the private sector, that ultimately bears the risk.

4   In the U.S., tax policies certainly influence privatization decisions. Because the interest on government bonds is tax exempt, many communities have strong financial incentives to use public rather than private capital. Asset sales may require the refinancing of tax-exempt debt with taxable debt, making it less attractive to investors and raising the cost of capital. Tax policies also work against the privatization of utilities that are exempt from corporate income taxes and property taxes. Tax exemptions, of course, shift rather than reduce costs: Although the costs of tax-exempt utilities are no lower, some of them are borne by federal or state taxpayers rather than by local ratepayers. Municipalities may be reluctant to give up such subsidies. Another financial cost associated with full privatization is a municipality's obligation to repay the un-depreciated portion of a government if it sells the facility that the grant supported. (Jerry Ellig, *The $7.7 Billion Mistake: Federal Barriers to State and Local Privatization*, Joint Economic Committee Staff Report, February 1996, pp. 7, 10, 11.)

Another factor discouraging the sale of wastewater treatment plants in the U.S. is the tougher environmental regulation that may result from them. The Resource Conservation and Recovery Act subjects the effluents from privately owned plants to higher – and more costly – standards than municipally owned plants must meet. According to the EPA, local governments indicate that the threat of being subject to new regulatory requirements is a "significant concern" when evaluating privatization options. (United States Environmental Protection Agency, *Response to Congress on Privatization of Wastewater Facilities*, EPA #832-R-97-001a, July 1997, pp. 5, 18, 22; and Ellig, *op. cit.*, p. 9.)

5   Disputes arose in four of the 16 water or wastewater utility privati-
    zations that Richard Dreese and Janice Beecher studied. Although
    the parties amicably resolved most of the disputes through arbitra-
    tion, litigation occurred in one case. (G. Richard Dreese and Janice
    A. Beecher, "To Privatize Or Not to Privatize: A POTW question for
    the 1990s," *Water Environment and Technology*, January 1997, p. 52.)
6   Joe Rogaly, "Hosing us for all we've got," *Financial Times*, July 12,
    1994.

INDEX

Abelson, Richard, 124

Aboriginal communities, water treatment in, 66-67

Accountability: in private sector, 13, 21, 134-36, 159; missing in public sector, 20; through enforceable contracts, 12, 24, 83, 133; through public monitoring, 29, 39. *See also* Freedom of information, Liability

Adam Smith Institute, 123, 124

Adams, Thomas, 155

Adcock, Robert and Natholyn, 129

Agostino, Dominic, 92

Agricultural pollution, 5, 6, 84

Alberta, privatization in, 81, 82, 84, 157

Alternative Financing and Public-Private Partnerships Working Group, 73, 76

Amalgamated Engineering and Electrical Union, 53, 121

American Federation of State, County, and Municipal Employees (AFSCME), 34, 38, 118, 120, 124, 125

American Water Works, 21, 97-98, 104

Anderson, John, 102, 108, 112, 113

Appropriations Committee of the House of Representatives, encouragement of privatization, 23

Aqua Alliance, 28

Asset sales: advantages of, 159; anticipated in Ontario, 74-75; discouraged by regulation, 153-54; discouraged in Ontario, 77; discouraged in U.S., 221n.4; in U.S. 17, 19, 26; opposed by Canadians, 42, 79; rare in Ontario, 75; rejected in Atlanta, 32; windfalls from, 20, 71; workers' share purchases, 123-24. *See also* England and Wales, Franklin, Ohio, Tax policies

Atlanta, Georgia, 18, 31-36; competitive selection, 32-33; cost savings, 26, 31, 33, 34; labour relations, 34, 123, O&M contract, 33; opposition to privatization, 32; performance disputes, 35-36; private investment in, 24, 33; rate increase moderated, 27, 31, 32; sewage pollution, 31; sewage privatization, 35, 36; socio-economic factors, 33; staffing levels, 34, 121; transition to private, 34-35; water quality, 35, 36

Azurix North America, 21, 36, 78, 82, 96-111, 136

Banff, Alberta, 84

Barham, John, 9

Barlow, Maude, 44

Barriers to privatization in Canada, 76-80, 160. *See also* Labour unions

Barron, Robert, 51

Beach closures: in Canada, 68; in England and Wales, 41, 46, 47; in France, 55, 59; in Hamilton, 91, 98; in U.S., 19

Bederman, Nolan, 110

Beecher, Janice, 153, 155

Benjamin, Daniel, 144

Benus, Arnold, 129

Bill 107 (Municipal Water and Sewage Services Transfer Act), 74, 77. *See also* Ontario Standing Committee

Bishop, William, 137, 139

Boil-water advisories, 5, 6-7, 36, 65, 67, 86, 88

Boucher, Paul, 77

Bragg, Ross, 86

Bribery. *See* Corruption

Briscoe, John, 11

British Columbia: boil-water advisories, 6-7, 65, 67; disease outbreaks, 67; initiative to involve private sector in wastewater treatment, 72; non-compliance of sewage treatment plants, 68, 130-31; privatization, 85-86; subsidies, 69, 70. *See also* specific municipalities

British Medical Association, 51, 52

Brockton, Massachusetts, 27

BTI Consulting Group, 21

Buenos Aires, 10, 11

Burlingame, California, 18

Bush, President, Executive Order by, 22

Byatt, Ian, 44, 50, 54

Campbell, Bill, 31, 32, 34, 35

Canada: experiments with privatization, 81-113; infrastructure crisis, 65-68

Canadian Council for Public-Private Partnerships (CCPPP), 77, 79, 81, 84, 117, 119, 126, 127

Canadian Environmental Defence Fund, 131

Canadian Environmental Law Association (CELA), 43, 44, 51, 78, 79

Canadian Union of Public Employees (CUPE), 43, 44, 49, 51, 53, 78, 79, 85, 86, 87, 91, 102, 105, 107, 108, 117, 118, 119, 121, 127

Canadian Water and Wastewater Association, 69, 71

Capital investment: amount required in Canada, 68-69, 70; amount required in U.S., 19; amount required world-wide, 11; as driver of privatization, 11, 20, 71, 86, 88; by private sector, 7, 9, 11, 24, 71, 72, 73,

82, 84, 88. *See also* Atlanta, England and Wales
Carson, Mac, 93, 94
Charlottetown, PEI, 68
China, 10
Churley, Marilyn, 43
Cincinnati, Ohio, 18, 221n.3
Clean Water Act, 18-19, 129; savings clause, 142
Clinton, President, Executive Order by, 23
Cobourg, Ontario, 143
Cochrane, Brian, 43
Cohen, David, 139
Collingwood, Ontario, 66
Combined sewer overflows: in Canada, 67, 68; in England and Wales, 45; in U.S., 19
Commodification of water, opposition to, 78
Common carriage. *See* Competition
Competition, 147-50, 156-57, 216n.3; common carriage, 148-49, 150; consumer choice, 147, 156; encourages efficiency and innovation, 12, 21, 95, 149, 156, 159; EPCOR as impediment to, 156; for Canadian contracts, 83, 84, 85, 86, 87, 88, 156; for U.S. contracts, 156; in capital market, 149; inset appointments, 148; limits to, 147, 150, 155; OCWA as impediment to, 77, 156; yardstick, 149. *See also* Hamilton
Compliance: as driver of privatization, 19, 20; improved compliance after privatization, 8, 28, 71, 72. *See also* England and Wales, Indianapolis, Non-compliance
Concession fees, 20, 26, 27
Concho Corporation, 129
Conference Board of Canada, 135
Conflicts of interest, as regulatory impediments, 7, 12, 108, 112-13, 130-34, 135, 136, 159, 160, 161. *See also* Enforcement, England and Wales, France
Contracts: components of, 12, 112-13, 133, 159, 160; number of, in U.S., 18; regulatory oversight of, 154-55; renewals, in U.S, 30
Cooke, Terry, 97, 100, 102, 105
Copps, Geraldine, 110
Coronado, Rick, 43
Corruption, 28-29, 32, 58, 154
Cory, Mr. Justice, 138
Cost of capital, 21, 76, 81-82, 87. *See also* Tax policies
Cost savings. *See* Efficiency, Savings from privatization
Cour des Comptes, 55, 58
Cowser, Brant, 40

Crane, Robert, 106

Crown immunity, 136-37; extended to contractors, 139; from consequences of policy making, 137-38; statutory protections from liability, 138-39

Cryptosporidium, 7, 66, 67, 85

Cuba, 10

Cunningham, Eric, 77

Cuomo, Mario, 24

CU Water Limited, 84

Darcy, Judy, 51, 78, 91

Daub, Mervin, 134

Davison, Michael, 92

Dawson City, Yukon, 67

Defence of statutory authority, 139-41; for Ontario sewage pollution, 140-41

Dickson, Mr. Justice, 137, 140

Dreese, Richard, 153

Drinking Water Inspectorate, 48, 150, 152

Earth Tech Canada, 84

Economic Policy Institute, 122

Economic regulation. *See* Regulation

Edmonton, Alberta (and environs), 84, 156, 157

Efficiency: of private sector, 7, 11, 12, 20, 21, 22-23, 24-25, 28, 57, 71, 72, 73, 85, 159, 168-69n.46; of public sector, 20-21, 168-69n.6. *See also* Competition, England and Wales, Hamilton, Price-cap regulation, Savings from privatization, Staffing levels

Eggers, William, 119, 123, 125, 135

Eisenberger, Fred, 97, 100, 102, 104, 111

Ellig, Jerry, 134

Elnaboulsi, Jihad, 57, 60

Enforcement: distance from abatement, 211n.12; improvement after privatization, 46-47, 103, 132-35; inadequate in Canada, 6, 7, 130-31. *See also* Conflicts of interest, Liability, Regulation

England and Wales, 41-54; beach closures, 41, 46, 47; capital investment, 41, 42, 44-45, 46, 48, 49, 152; compensation for poor performance, 48; competition, 147-50; conflicts of interest, 42, 132; customer service, 42, 48-49; disconnections, 50-51, 181n.74; disease, 51-52; drinking water quality, 42, 47-48, 150; economic regulation, 150-53; efficiencies moderating rate increases, 50, 217n.17; efficiencies questioned, 44-45; environmental performance, 42, 45-46; executive salaries, 50; myths perpetrated by labour and environmental groups, 42-44, 45, 49, 51; privatization process, 42, 177n.8; profits and dividends, 50, 152, 218n.23; rate increases and reductions, 49-50, 152-53; regulation and prosecution, 46-47, 48, 54, 132-33, 150; restructuring, 152-53, 219n.28; staffing levels, 53, 119, 121; water losses, 41, 46; windfall tax, 50, 152, 177n.8

Environment Agency, 45, 46, 47, 133, 150

Environmental performance. *See* Compliance

EPCOR, 156

European Community/Commission/Union, 41, 46, 55, 56, 57, 59, 150

Eves, Ernie, 75

Expertise in private sector, 7, 11, 12, 13, 21-22, 25, 28, 36-37, 56, 72, 73, 83, 88, 160

Export: of water, 78; of water services, 53, 71, 72, 73, 97

Fantauzzo, Stephen, 39

Federation of Canadian Municipalities, 65, 70, 81, 117

Ferguson, Mark, 85

Filer Consultants, 98

France, 55-61; beach closures, 55, 59; conflicts of interest, 60; corruption, 58; drinking water quality, 56, 58-59; entry fees, 57, 58; history and extent of privatization, 55, 56; limited competition, 58; public ownership of assets, 55, 56; rate increases, 57, 58, 184n.23; sewage treatment and pollution, 55-56, 59-60; shortcomings of public provision, 55; subsidies, 56-57, 60; water losses, 56

Franklin, Ohio, 17, 27, 28

Franklin, Shirley, 35-36

Freedom of information, 113, 136, 155. *See also* Transparency

Gallagher, John, 113

Galt, Doug, 69

Georgia House Bill 1163, 23

Goderich, Ontario, 83-84, 156

Gohier, Leo, 93, 100, 102, 107, 108

Goldsmith, Stephen, 37, 38, 39, 40

Grubel, Herbert, 125

Haldimand-Norfolk, 82

Halifax, Nova Scotia: privatization process, 86-87, 119, 156; savings, 87; sewage pollution, 67, 86, 87

Hamilton, Ontario, 82, 91-113; access to information, 109, 113; beach closings, 91, 98; capital investment, 91, 92-93, 97, 111; competition in future, 111; disputes over savings, 106-7; economic development, 91, 92-93, 94, 95, 96-98, 111, 112; efficiencies, 105-6; enforcement, 102-3, 112; exemptions from liability, 101-2, 107, 112; guarantees, 95-96; labour relations, 96, 105, 126; problems with public operations, 91-92; responsibility for capital improvements, 107, 112; restraints on public criticism, 108, 112-13; savings, 91, 94, 95, 103-6; sewage pollution, 91, 98-103, 110; sole-sourcing, 94-95, 110-11, 157; staffing levels, 97, 105, 107, 119, 202n.113

Hayek, Friedrich, 131-32
Hayman, Baroness, 47
Hoath, Greg, 105, 107, 108, 126
Hoboken, New Jersey, 24
Houston, Texas, 18, 25, 28
Hudson Institute, 19, 20, 24, 26, 27, 28
Hunt, Marvin, 85

Incentives, for private firms, 12, 21, 28, 38, 45, 71, 73, 104, 105, 107, 112, 130, 150. *See also* Liability, Price-cap regulation
Indianapolis, Indiana, 18, 21, 29, 36-40; competitive selection, 36; cost savings, 36, 37, 38; environmental performance, 36, 39-40; labour relations, 33, 38-39, 123, 124, 125-26; managed competition, 220n.2; rate increases avoided, 37; staffing levels, 37, 119, 121; workplace safety, 38, 124
Industry Canada, 72
Innovation in private sector, 7, 12, 21-22, 25, 71, 72, 73, 95, 149
Inset appointments. *See* Competition
International trade agreements, 78-79, 85, 86
International Union of Operating Engineers (IUOE), 96, 99, 102, 105, 107, 126
Investment. *See* Capital investment

Joint Economic Committee of Congress, 20, 134

Kamloops, BC, 85-86
Kelowna, BC: cryptosporidium outbreak 67; water meter contract, 81
Kinnersley, David, 41, 42, 132
Kitchener-Waterloo, Ontario, cryptosporidium outbreak, 66
Kitchen, Harry, 25
Klein, Michael, 76, 147, 219nn.28, 31
Koebel, Frank and Stan, 5-6
KPMG, 94

Laborers' International Union of North America, 126

Labour unions, 117-27; as obstacle to privatization, 117; opposition to privatization, 8, 53, 78, 118-19, 125, 127; representing private employees, 124, 127; support of privatization, 124, 125-26, 127. *See also* Asset sales, Atlanta, England and Wales, Hamilton, Indianapolis, specific unions, Staffing levels, Training, Wages and benefits, Workplace safety

La Forest, Mr. Justice, 140

Laughren, Floyd, 43, 45, 74

Leitch, Art, 91

Liability, 83, 84, 129-45; as incentive to reduce risk, 130, 136, 139, 140, 141, 142, 144, 210n.7, 214n.53; barriers under public law, 130-36; barriers under tort law, 136-41; contractors' unwillingness to accept, 99, 102, 107, 144-45; personal liability of owners, 129; restoring, 141-45; under tort law, 143-45. *See also* Accountability, Crown immunity, Defence of statutory authority, Regulation, Risk

Linden, Allen, 136

Lipson, Will, 94

Littlechild, Stephen, 150, 151

Livernois, John, 6

London, Ontario, 83

Loxley, John and Salim, 107, 108, 109

Luross, Bob, 84

MacDonald, Jack, 93

Major, John, 130

Managed competition, 160, 220n.2, 221n.3

Mancinelli, Joseph, 126

Manila, 10, 11

Marine Conservation Society, 46

Martin, Amanda, 78

May, Janet, 43

McIntyre, Jeff, 98

McManus, Pat, 21

Meacher, Michael, 46, 47

Meiners, Roger, 141

Miller, Sarah, 43

Milwaukee, Wisconsin, 18, 26, 27, 28, 133, 205n.162; cryptosporidium outbreak, 66; labour relations, 33, 119, 121, 123, 124, workplace safety, 124

Mitchell, Anne, 51

Moncton, New Brunswick, 87-90; drinking water quality, 88, 89; financing, 82, 88; privatization process for filtration, 88; privatization process for network, 89-90; savings, 88-89

Monopoly. *See* Competition, Regulation

Montreal, Quebec: primary sewage treatment, 68; thoughts of privatizing, 81; water losses, 65

Moore, Cherie, 38-39

Morra, Sam, 69

Municipalization, in U.S., 29-30

Munro, John, 111

Murphy, Brian, 89

Murphy, John, 126

National Association of Water Companies, 24

National Commission on Employment Policy, 119, 121

National Regulatory Research Institute (NRRI), 17, 20, 153, 154, 155

National Round Table on the Environment and the Economy, 9, 69, 71-72

New Brunswick, subsidies, 69. *See also* Moncton, Saint John

Newfoundland: boil-water advisories, 6, 65, 67; subsidies, 69; trihalomethanes in reservoirs, 67. *See also* St. John's

New Orleans, Louisiana, 18, 25, 28

Non-compliance, with U.S. regulations, 18-19. *See also* Compliance, Liability, specific municipalities and provinces

North Battleford, Saskatchewan, 7

Nova Scotia: non-compliance with regulations, 68; promotion of privatization, 72, 86; subsidies, 69; trihalomethanes in reservoirs, 67. *See also* Halifax

O'Bannon, Frank, 39

Ofwat, 46, 48, 50, 52, 148, 149, 151, 152, 153

OMI / OMI Canada, 36, 84

Ontario: artificially low prices, 70-71; boil-water advisories, 6, 65; disease outbreaks, 5-6, 66; government inaction on privatization, 74-75; government support of privatization, 73-76, 75; inadequate training of operators, 66; non-compliance of sewage treatment plants, 68, 131; non-compliance of water treatment plants, 6, 65, 66, 131; O&M contracts, 82-84, 91-113; shortage of funds, 70; subsidies, 69, 70, 77, 190n.56

Ontario Clean Water Agency (OCWA), 74, 75, 77-78, 83, 131, 156

Ontario Ministry of the Environment, 6, 66, 74, 99, 102, 131

Ontario Public Service Employees Union (OPSEU), 44, 51

Ontario Standing Committee on Resources Development, hearings into Bill 107, 43, 44, 45, 49, 51, 53, 78, 153

Organisation for Economic Co-operation and Development (OECD), 60, 71

Pacheco Law, 122-23
Patronage, 20-21, 125
Paver, Paul, 129
Payment, Pierre, 66
Performance. *See* Compliance, Efficiency, Savings, specific jurisdictions
Performance payments and penalties. *See* Incentives
Philip Utilities Management Corporation / Philip Environmental / Philip
    Services, 77, 92-111, 133, 153
Pirie, Madsen, 124, 126
Poole, Robert, 117, 134-35
Power, Michael, 69-70
Powers, Ross, 110
Power Workers' Union, 126
Pratt, Glenn, 39, 40
President's Commission on Privatization, 22
Price-cap regulation. *See* Regulation
Price Waterhouse, 73, 86
Pricing: artificially low in Canada, 70-71; de-politicization of, 12, 154,
    155; privatization to align prices and costs, 12, 24, 71, 72; rate increas-
    es required to meet investment needs, 71; rate increases tempered by
    privatization, 27; to promote conservation, 12, 70. *See also* Atlanta,
    England and Wales, France, Indianapolis, Subsidies
Privatization, extent: in BC, 85; in Canada, 81; in France, 55, 183n.4; in
    Ontario, 82; in U.S., 17-18; worldwide, 9-10. *See also* specific jurisdic-
    tions
Professional Services Group (PSG), 28-29, 82
Provincial-Municipal Investment Planning and Financing Mechanism
    Working Group, 73
Public Services International Research Unit (PSIRU), 118, 127
Public Works Financing, 9, 18, 25, 30, 122, 123

Quebec: boil-water advisories, 6, 65; Commission on Water Management,
    71; disease outbreaks, 66; inadequate monitoring, 68; sewage pollu-
    tion, 68; subsidies, 69. *See also* Montreal
Quick, Steve, 38, 124, 125

Rafferty, Liam, 94
Rate-of-return regulation. *See* Regulation
Rates. *See* Pricing
Reason Foundation, 9, 25, 27, 53, 117, 119, 122, 134, 135
Regulation: as barrier to full privatization, 152-54, 221n.4; citizen
    involvement in, 155, 161; economic, 150-55; importance of, 8, 12, 132,

134, 150, 154, 160-61; inadequate in Canada, 186-87n.7; limitations of, 136, 142, 144, 155-56; of contracts, 154-55; price-cap, 150-52; principles to guide, 155; rate-of-return, 150, 154, 217n.16; supplementing tort law, 143-45. *See also* Conflicts of interest, Enforcement, England and Wales, Liability

Renzetti, Steven, 70

Reputation, importance of, 21, 120, 135-36

Risk, transfer to private sector, 11, 72, 73, 76, 82, 86, 87, 112, 144, 160. *See also* Liability

Rogaly, Joe, 161

Ryan, Sid, 43, 51, 53, 78

Sabbaghi, Asghar, 147

Safe Drinking Water Act, 18, 19, 129

Saint John, New Brunswick, sewage pollution, 67

St. John's, Newfoundland: liability limitation, 139; sewage pollution, 67

Sanitation: number of people without, 11; deaths related to, 11

Saskatchewan, deficient water systems, 7, 65

Savings from privatization, 21, 24-26, 27, 72, 74, 81, 83, 84, 85, 87, 88, 123, 148, 151, 159. *See also* Atlanta, England and Wales, Hamilton, Indianapolis

Sawicki, Joan, 68

Seattle, Washington, 18, 26, 96, 144

Serageldin, Ismail, 11

Sewage pollution: in Canada, 7, 67-68, 83, 86-87, 91, 98-103, 110, 131, 140-41, 143; in developing world, 11; in England and Wales, 41, 45-47; in France, 55-56, 59-60, 61; in U.S., 18-19, 23, 31, 39, 129

Shavell, Steven, 144

Shellfish closures, 68

Shewfelt, Delbert, 83

Shrybman, Steven, legal opinion, 79

Sierra Legal Defence Fund, 67, 68, 186n.7

Smith, Stuart, 77, 93-94, 95, 96, 97, 98, 99, 103, 133, 153

Sole sourcing, 156-57. *See also* Hamilton

Sorger, George, 99-100

South Bay Utilities, 129

Spencer, Stan, 108

Spulber, Nicolas, 147

Staffing levels: in private utilities, 25, 83, 118, 119-20; in public utilities, 11-12, 21, 125; protection from layoffs, 120-21. *See also* Atlanta, England and Wales, Indianapolis, Labour unions

Stayton, Mike, 37

Steele, John, 102
Sterling, Norman, 74, 75, 134
Stitt, Skip, 20, 22, 25, 120, 123
Stokes, John, 21, 101, 136
Strang, Al, 88, 89
Stroup, Richard, 141
Subsidies: in Canada, 69-70, 76, 77, 82, 87, 131; in developing countries, 12; in France, 56-57, 58, 60, 61; in U.S., 20, 28, 135, 221n.4; from liability exemptions, 141. *See also* Pricing
Suez Lyonnaise des Eaux, 10, 11, 21, 33, 36, 55, 82, 87
SuperBuild Corporation, 70, 74, 75
Swain, Harry, 119

Tabuns, Peter, 44, 51, 75
Tampa / Tampa Bay, Florida, 18, 26
Tax policies, influence on privatization, 17, 22, 29, 33, 76, 77, 152, 156, 160, 221n.4
Tennessy, Tony, 126
Termination provisions, 83, 84, 87, 96, 133
Thompson Gow and Associates, 65, 71
Tort law. *See* Liability
Training, 5, 6, 13, 20, 34, 35, 38, 53, 65, 66, 93, 105, 125, 126, 186n.7
Transparency, 29, 89, 155. *See also* Freedom of information
Trebilcock, Michael, 110, 130, 142
Trihalomethanes, 35, 67, 89

United States: asset sales, 17, 19, 26; contracting out, 17-40; extent of private ownership, 17
United States Environmental Protection Agency, 17, 18, 19, 27, 40, 129; proponent of privatization, 23
United States General Accounting Office, 19, 24, 120
United Water / United Water Services Canada, 21, 24, 33, 36, 77, 82, 87, 110. *See also* United Water Services Atlanta, White River Environmental Partnership
United Water Services Atlanta, 33-36
USFilter / USF Canada, 21, 28, 29, 35, 82, 83, 87-90, 125
U.S. Public Interest Research Group, 19

Vancouver, BC / GVRD: non-compliance of sewage treatment plants, 68; primary sewage treatment, 68; Seymour water plant, 85, 86, 119, 144
Van Frank, Richard, 40
Victoria, BC: septage contract, 85; sewage pollution, 67

Vietnam, 10
Vivendi, 10, 21, 28, 55, 82, 83, 88

Wages and benefits, 34, 83, 118, 121-23, 127
Walkerton: Inquiry, 74, 75, 104, 134, 155; water tragedy, 5-7, 8, 65, 70, 84, 131
Wallace, Larry, 34
Warren, Gil, 43, 75
Wastewater pollution. *See* Sewage pollution
Water Infrastructure Network, 19, 21
Water losses, 11; in Atlanta, 36; in Canada, 65, in England and Wales, 41, 46
Water supply: number of people without, 10; deaths related to, 11
Water use: in Canada, 70; in Ontario, 71
Water Watch, 43, 49
Wheeler, David, 49
White River Environmental Partnership, 36-40
White Rock Utilities, 85
Williamson, Michael, 53
Williams-Russel and Johnson, 33
Wilson, Dave, 92, 93, 94, 104, 109-10
Wilson, Madam Justice, 138
Winfield, Mark, 51
Winnipeg, Manitoba, 81
Winter, Ralph, 142
Wolf, Tom, 124
Workplace safety, 36, 38, 124
World Bank, 9, 11, 12, 123
World Commission of Water for the 21st Century, 11
World Water and Environmental Engineering, 9-10
World Water Council, 11

Yemen, Peter, 96, 105, 107

ACKNOWLEDGEMENTS

This book owes much to the research conducted by several colleagues over the last four years. I am indebted to Craig Golding for his papers on France and Indianapolis; Alex Orwin for his survey of privatizations around the world; Krystyn Tully for her research on Atlanta; Doug Paisley for his work on Hamilton and the Ontario Clean Water Agency; and Tom Adams for his insights into utility regulation. Geoffrey Patridge spent countless hours at the library on my behalf; Lisa Peryman and Kevin Lacey tracked down many an elusive fact; Andrea Kourato translated French documents; and Tom Richardson pored over hundreds of news stories. The book also benefitted from several excellent editors: Andrew Stark, Larry Solomon, and two anonymous reviewers. And I am grateful to my husband, Richard Owens, not only for providing reference materials on demand but also for engaging in endless conversation about Canada's ailing utilities, despite his reservations about discussing sewage at the dinner table.

This book would not have been possible without the financial support of Environment Probe's individual donors and grants from several charitable foundations. I heartily thank the Helen McCrea Peacock Foundation, the Donner Canadian Foundation, and the Earhart Foundation, whose support was facilitated by the Political Economy Research Center.